Reparations and the Human

Reparations and the Human

David L. Eng

Duke University Press *Durham and London* 2025

Printed in the United States of America on acid-free paper ∞
Project Editor: Liz Smith
Cover designed by A. Mattson Gallagher
Typeset in Portrait by Copperline Book Services

Library of Congress Cataloging-in-Publication Data
Names: Eng, David L., [date] author.
Title: Reparations and the human / David L. Eng.
Description: Durham : Duke University Press, 2025. | Includes
bibliographical references and index.
Identifiers: LCCN 2024039433 (print)
LCCN 2024039434 (ebook)
ISBN 9781478031864 (paperback)
ISBN 9781478028628 (hardcover)
ISBN 9781478060840 (ebook)
Subjects: LCSH: Reparations for historical injustices. | Human rights. |
Colonies—Psychological aspects. | Decolonization—Social aspects. | Social
justice—Environmental aspects. | Social justice—Psychological aspects.
Classification: LCC KZ6785 .E54 2025 (print) | LCC KZ6785 (ebook) |
DDC 342.08/8—dc23/eng/20241008
LC record available at https://lccn.loc.gov/2024039433
LC ebook record available at https://lccn.loc.gov/2024039434

Cover art: Photograph by the author.

publication supported by a grant from
The Community Foundation for Greater New Haven
as part of the Urban Haven Project

Contents

Preface and Acknowledgments

Most of the world's uranium supply is mined from Indigenous lands. The uranium used to produce the only two nuclear weapons ever deployed on human populations, those dropped on Hiroshima and Nagasaki during World War II, came in part from the territories of the Sahtu Dene, a First Nations people inhabiting northwestern Canada. Many of the Sahtu Dene men who labored on behalf of this atomic initiative died of cancer. To this day, their descendants suffer from extraordinary rates of malignancy and premature death. Although the Canadian government shuttered the mine in 1960, the land on which the Sahtu Dene live and the ecosystem on which they depend for their sustenance is poisoned by radioactive waste for eternity. Ignorant at the time of how their mining efforts would be applied or the destination of the ill-fated ore, the Sahtu Dene nonetheless felt implicated once they learned about their connection to those who were annihilated in Japan. In response, they sent a delegation to Hiroshima to apologize.

There is little in Western political or psychoanalytic theory that can adequately account for this extraordinary act and the Sahtu Dene's attempts to repair the violence of atomic destruction in Hiroshima for which they are, arguably, the least responsible. To be sure, Western conceptions of the sovereign state and the liberal individual underpin long histories of European colonization and trauma that seek to police the boundaries of the human being as well as the limits of repair in the face of unremitting war and violence. *Reparations and the Human* interrogates political and psychic genealogies producing "bad" colonial subjects outside the pale of reparations and eccentric to definitions of the human. A category whose shifting values under colonial modernity have created clear distinctions among the very aggregate it seeks to universalize, "the human" is a comparatively recent historical term. The nominalization of an adjective—as in "human being"—it first appeared in 1840, according to the *Oxford English Dictionary*.[1] This book interrogates normative theories of the liberal human—and of repair, trauma, and sovereignty—evolving under European colonialism. However, it also necessarily moves *beyond* them to offer a different account of reparations and the human—from the time of colonial conquest in the New World to the era of nuclear devastation in Asia to the contemporary moment of the Sahtu Dene's surprising actions in Hiroshima. In other words, *Reparations and the Human* seeks to do justice to those subjects injured yet long unacknowledged by conventional distinctions between victims and perpetrators in Western law, politics, and theory.

THIS BOOK REPRESENTS an expansion of my prior scholarship on law, psychoanalysis, and the Transpacific insofar as I have been compelled to engage with several fields and disciplines—from Indigenous and environmental studies to war history and diplomatic relations to scientific processes of uranium mining and enrichment—that required extended research, reading, contemplation, and synthesis. In this regard, I have many friends, colleagues, institutes, and foundations to thank for their continuous tutelage, support, and inspiration over the decade it has taken me to complete this book.

This project began with my prior writings on transnational adoption in Asia and the psychic possibilities of racial reparation. However, it blossomed into something more ambitious and significant in 2012–13, when I spent a year as a member of the School of Social Science at the Institute for Advanced Study (IAS) in Princeton. It is a small irony that the IAS, which the "father of the atomic bomb" J. Robert Oppenheimer directed from 1947 to 1966, became the place where I began to contemplate systematically how reparation converged and diverged across the political and psychic domains. I am especially grateful to André Dombrowski, Mark Driscoll, Neve Gordon, Moon-Kie Jung, Maria Loh, the late Diane Nelson, Nicola Perugini, Catherine Rottenberg, and Joan Scott—as well as Danielle Allen, Nicola Di Cosmo, Nancy Cotterman, the late Freeman Dyson, Sara Farris, Didier Fassin, Donne Petito, Judith Surkis, Peter Thomas, and Heidi Voskuhl—for creating an inimitable atmosphere of intellectual and social conviviality in the most improbable town in New Jersey. I am equally grateful to the Helsinki Collegium for Advanced Studies, and its then director Sami Pihlström, for providing me with an additional year of research and writing time in 2015–16 to develop this book.

Many close friends and colleagues have been steadfast interlocutors and supporters of this project over the years, offering keen insights and interventions alike: Tony Anghie, Emma Bianchi, Jodi Byrd, Zahid Chaudhary, Patricia Clough, the late Christina Crosby, Cathy Davidson, Gina Dent, Ezekiel Dixon-Román, Ivan Drpić, Stuart Freeman, Tak Fujitani, Shinhee Han, Janet Jakobsen, the late Amy Kaplan, Jonathan Katz, Suvir Kaul, Fariha Khan, David Young Kim, Homay King, Hali Lee, Susie Lee, Eng-Beng Lim, Ania Loomba, Jerry Miller, Susette Min, Mae Ngai, Naomi Paik, Jennifer Ponce de León, Jasbir Puar, Camille Robcis, David Seitz, Lara Sheehi, Stephen Sheehi, Shu-mei Shih, Kaja Silverman, Michelle Stephens, Priscilla Wald, Chi-ming Yang, Lisa Yoneyama, and Hiro Yoshikawa. Shortly after I completed what I thought was the final draft of this book, my long-standing, entertaining, perspicacious, and very opinionated NYC-via-CA crew—Ed Cohen, David Kazanjian, Teemu Ruskola, Josie Saldaña, and Melissa Sanchez—convened a manuscript workshop at my behest. The conversation not only gave rise to the concept of the "beyond" now organizing this book but also sharpened its critical arguments in important and unanticipated ways. During several very unfun pandemic years, I was fortunate to work with a very fun and equally astute group of doctoral students at the University of Pennsylvania, all treasured colleagues in the profession now. Melanie Abeygunawardana, Matthew Aiello, Thomas Conners, and Ava Kim read portions of the manuscript in our Dissertation Working Group and provided a steady stream of fantastic feedback.

I am sustained by extraordinarily talented colleagues at Penn—especially in the Department of English and the Program in Asian American Studies—whose scholarly depth, institutional commitment, and dedication to our students make it inspiring to show up for class. Thank you, Hamit Arvas, Ericka Beckman, Pearl Brilmyer, Rita Copeland, Margo Crawford, Hardeep Dhillon, Anabel Bernal Estrada, Michael Hanchard, Nancy Hirschmann, John Jackson, Melissa Jensen, Fariha Khan, Baki Mani, Emily Ng, Ann Norton, Zita Nunes, Rupa Pillai, Anne Marie Pitts, Kevin Platt, Jean-Michel Rabaté, Paul Saint-Amour, Deb Thomas, Filippo Trentin, Loretta Turner, David Wallace, and Dag Woubshet, for being who you are.

Parts of *Reparation and the Human* have appeared in previously published forms. A section from chapter 1 appeared as "Colonial Object Relations" in *Social Text* and as "Reparations and the Human" in the *Columbia Journal of Gender and Law*; a section of chapter 2 appeared as "The History of the Subject and the Subject of History" in *History of the Present*; and a section of chapter 3 appeared as "Reparations and the Human" in *MLA Profession*.

Finally, I would like to thank A. Mattson Gallagher, Kate Mullen, Chris Robinson, Chad Royal, and Liz Smith at Duke University Press for shepherding this book to completion and Matthew MacLellan for preparing the index. I could not imagine publishing *Reparations and the Human* with anyone other than the inestimable Ken Wissoker, who has been my stalwart friend and editor at Duke for three decades and four monographs strong—or writing it for that matter while being with anyone other than my brilliant and patient Moomin partner, Teemu Heikki Petteri Ruskola.

Introduction

Two catastrophes in the twentieth century marked a radical shift in our conceptions of the human being and, more specifically, of human precarity. The Holocaust and the atomic bombings of Hiroshima and Nagasaki invoked in graphic terms the specter of total human destruction. In *The Origins of Totalitarianism*, Hannah Arendt famously wrote that citizenship is nothing more or less than "the right to have rights," and she observed in the context of European warfare and its crises of statelessness that "the world found nothing sacred in the abstract nakedness of being human."[1] By deriving the human from its inscription in Western law, the legal status of an individual under European modernity and sovereignty came to predicate rather than to describe all human beings and their fundamental rights in a long and uneven history of colonial settlement and violence. In time, the skeletal figures liberated by Allied troops from German death camps and the suppurated frames of survivors of nuclear disaster in Japan stretched not only Arendt's trenchant analyses of totalitarianism but also the human imagination itself to index a realm of bare life altogether beyond rights—of "life unworthy of life" (*lebensunwertes Leben*)—and those with no right to live at all.

In response, a new international order of human rights with attendant notions of reparations arose from the ruins of World War II. This new legal regime sought to subrogate the sovereignty of the nation-state in order to defend the sovereignty of the human being—however long the latter had been subordinated to the former concept in Western law and politics. Traditionally, reparations could be claimed by a victorious nation-state from a defeated one as compensation for the costs of war. For the first time in history, reparations were extended to encompass individual and group claims against state-sponsored violence and crimes against humanity. Confounding prior legal divisions separating public war from private grievance, as well as state responsibility from civil liability, reparations and human rights sought to protect the abstract nakedness of being human and to compensate for the striking failures of the modern nation-state to ensure the sanctity of human life—a new dream of enlightenment emerging from the ashes. Great evils of the past and present, as the law scholar Martha Minow observes, are nothing new. What was new were "mounting waves of objections and calls for collective responses to mass violence."[2]

While genocide in Europe and nuclear holocaust in Asia oriented the human imagination toward the specter of planetary annihilation—a prospect accelerated by the advent of Cold War hostilities between East and West under the threat of "mutual assured destruction"—the "final solution" and the atomic bombings also cleave from one another in significant ways. In the space of postwar Europe, the history of the Holocaust is settled: Nazis were perpetrators and Jews were victims. In contrast, in the space of postwar Asia, there was and continues to be little historical consensus as to who were the victims and who were the perpetrators in the aftermath of atomic destruction. Unlike the Holocaust, whereby Germany paid reparations to Jews as well as to the state of Israel, the possibility of reparations for those who survived the atomic bombings remains unthinkable. Indeed, nuclear weapons to this day remain legal instruments of warfare under international law such that "sovereignty *is* nuclear weapons," in the words of international legal theorist Antony T. Anghie.[3] My book begins with this notable disjuncture to explore a history of reparations and the human in Cold War Asia.

Reparation is a key term in both political theory and psychoanalysis—particularly in object relations—but the concept functions very differently in each field and is rarely discussed across them. While political theory conceives of reparations primarily as a noun—an event, a response, an accounting, a payment meant to write a history of violence into the past—psychoanalysis approaches the concept more as a verb: a continuous process mediating relations between love and hate, between idealization and aggression, among contending subjects. We should neither bemoan this slippage nor consider it a political liability. To the contrary, the gap between *reparations* (in the plural) as a noun and *making reparation* (in the singular) as a verb keeps open a space for new victims to be apprehended and new injuries to be claimed. In other words, as psychoanalysis indexes the radical indeterminacy of human relations—including the desire to repair and thus the will to maintain a relationship to an injured other—it becomes a vital heuristic for exploring the social and psychic effects of power, violence, and the politics of redress. These overlapping processes establish the conditions of possibility for the emergence of the human being in the first instance while also subordinating subjects and populations deemed less than human.

It is important to emphasize from the outset that Arendt's stateless and displaced persons in Europe were hardly the first group to suffer the contingencies of being human. Those who were colonized and enslaved before them had given the lie to any fundamental concept of natural or universal rights long before the traumas of Hitler's Third Reich. A distinguished line of Afro-diasporic and African American thinkers from Frantz Fanon and Sylvia Wynter to Hortense Spillers, Alexander G. Weheliye, and Zakiyyah Iman Jackson have interrogated the social and the psychic conditions under which one qualifies for inclusion as a full member of the human race.[4] Given the brutal history of European colonization and settlement, along with the ravages of the transatlantic slave trade and its enduring legacies, Fanon concluded that the project of decolonization would require nothing less than "quite simply the substitution of one 'species' of mankind by another."[5]

Reparations and the Human extends the critical insights of this intellectual tradition in Black studies to investigate the distinct forms this problematic takes in the space of the Transpacific. I offer a novel genealogy of reparations and the human in Western political and psychoanalytic thought by insistently placing the concepts in a comparative historical and geographical context. The book considers relations among Europe, the Americas, and Asia from the Age of Discovery (chapter 1) to the aftermath of genocide in Germany and nuclear holocaust in Japan (chapter 2) to our present moment of atomic fallout and environmental collapse (chapter 3). It analyzes a long history of colonial modernity and how, in turn, political and psychic genealogies of reparation established during the settlement of the Americas and the rise of European Enlightenment continue to determine shifting configurations of the human being and human rights in the face of war and violence across the Cold War Transpacific.

Throughout the book, I track how the idea of reparations appears as a central concept mediating sanctioned as well as disavowed histories of human trauma and suffering. The acknowledgment of trauma and suffering and the will to repair retroactively confer the juridical categories of victim and perpetrator, along with psychic categories of good and bad objects, colonizing and Indigenous subjects—deciding who is worthy or unworthy of redress. Together these social and psychic processes delimit the notion of the human being across disparate times and spaces. This book traces the dynamics structuring the differential productions of this figure—its arrogation of injury, its claims to political recognition, its rights to economic compensation.

As Western conceptions of the sovereign state and individual constitute the political foundations for European colonization and conquest by demarcating the boundaries of the liberal human and its rights to repair, my book necessarily investigates normative theories of repair, trauma, and sovereignty on the collective and individual level, and across the social and psychic domains. However, I necessarily move *beyond* these foundational categories to offer another account of reparations and the human—an alternative approach to repair and responsibility in a precarious world of vulnerable subjects. The philosopher Elizabeth V. Spelman observes that the act of repair is a core aspect of human life. We live in a world full of violence and aggression, one constantly in need of fixing. Yet we repair only that which is valuable to us.[6] *Reparations and the Human* takes as its point of departure this simple insight to consider who and what is considered deserving of attention and care, of repair.

Chapter 1, "Beyond Repair: Political and Psychic Genealogies in Locke and Klein," traces a conceptual history of reparations from the political theories of John Locke (1632–1704) to the psychoanalytic theories of Melanie Klein (1882–1960), beginning with dispossession and death in the New World and ending with the nightmare of total war and genocide in the Old World. *Reparation* in Locke becomes a key political term for establishing the natural rights of sovereign European man in liberal political theory while simultaneously functioning to justify a differential redistribution of life, liberty, and property across the Transatlantic. In the space of Locke's New World, the concept of reparations does not limit violence and restitution as it does in Europe. To the contrary, it comes to rationalize a political process by which genocide and an unfettered appropriation of Indigenous lands can be pursued in the name of justice.

While there is little scholarship in political theory focusing on the limits of reparation in response to claims of colonial injury, there is even less critical attention paid to the colonial dimensions of object relations or, more generally, to the psychic dimensions of liberal reason. Concomitantly, I explore in chapter 1 how reparation in Klein functions as a key psychoanalytic concept producing a closed circuit of victims and perpetrators. Klein developed her notions of reparation in the interwar years, as the dream of enlightenment dissolved into ashes, but she embedded her theory in an earlier scene of European war and settlement. Nearly 250 years after Locke, Klein propounds a theory of psychic violence and repair, one implicitly analyzing the legal, political, and economic effects of English—and European—colonizing violence that Locke's philosophical writings sought to legitimate.

In Klein, reparation constitutes the colonial settler as *both* an aggressive perpetrator *and* a defensive victim. On the one hand, by arrogating violence and trauma for the self-same European subject, Klein's notions of reparation produce the colonial settler as a good object deserving of repair; on the other hand, they simultaneously displace and manage a brutal history of aggression against Native populations through a colonial morality that configures the Indigenous other as a bad object undeserving of any human consideration. I describe this differential psychic production of repair as "colonial object relations."

Colonial object relations are not a moral response to violence but, rather, an instrumental effect of colonial reason. They condition the possibilities and limits of repair by establishing a politics of recognition, identity, and injury—of who is deserving and what is capable of being redressed—under European sovereignty and liberalism, determining its uneven distribution of universal rights and its lethal dispensations of the human and human life. In contrast to a number of recent psychoanalytic commentators, I suggest that reparation in Klein offers more than just a theory about the salvaging of a relational tie between contending subjects of violence. It also delineates a psychic practice—one established precisely through an unresolved history of colonial conquest, racial capitalism, and their biopolitical legacies—by which violence is preserved *in potentia* for the consolidation of liberal subjectivity and nation-building. Ultimately, the limits of *racial* reparation index the social and psychic dilemma of repairing not the good but the bad objects of colonial conflict produced by a long history of Enlightenment bad faith. In the same breath, it calls for a postcolonial critique of psychoanalysis.

Placing Locke and Klein in conversation with one another reveals an economy of idealization and aggression—a "coeval violence of affirmation and forgetting"—that, in the words of postcolonial scholar Lisa Lowe, "structures and formalizes humanism. This economy civilizes and develops freedom for 'man' in modern Europe, while relegating others to geographical and temporal spaces that are constituted as uncivilized and unfree."[7] Simultaneously, it demonstrates how the social contract is also, and indeed, a psychic contract—a psychic as well as racial contract, to extend the writings of political theorist Charles W. Mills in an alternative direction.[8] From another perspective, the juxtaposition of Locke and Klein resituates political theorist Wendy Brown's critique of the codification of "states of injury" and the protocols of ressentiment in contemporary US identity politics by relocating its proper subject not in the racially subordinate but, rather, in a longer genealogy of the European liberal human and *its* injured history of consciousness.[9] Investigating the psychic dimensions of this privileged subject reveals how liberalism and colonialism remain indissociable even when liberalism seeks to repair the injuries of colonialism and even when it seeks to redistribute property in the name of justice. In the final analysis, Locke and Klein underscore how political and psychic processes of reparation are profoundly compromised in the history of liberal reason—indeed, how they ultimately expose a problem of *racial* reparations and the human. Chapter 1 thus offers an alternative account of repair from the perspective of the colonized rather the colonizer.

Whereas chapter 1 explores reparations through the problem of *colonization* and the rights of man in the Americas, chapter 2, "Beyond Trauma: War and Violence in the Transpacific," examines the concept in terms of *decolonization* and human rights in Cold War Asia, as the world was forced to confront new biopolitical technologies of violence threatening to annihilate planetary life.[10] The modern concept of genocide emerged in the wake of the Holocaust in Europe, while nuclear devastation in Japan inaugurated the Atomic Age. In the same moment, the atomic bombings connected the specter of nuclear holocaust singularly and indelibly to Asia. Today we imagine atomic destruction in the language of "nuclear universalism," one threatening the existence of every living creature and thing on planet Earth, yet the Asian origins of "ground zero" must not be forgotten.

In the catastrophic wake of the Holocaust and atomic bombings, the liberal rights of European man were, as Arendt underscored, both inadequate and exhausted. Recast and reinvented as universal human rights under the shadows of Cold War decolonization, the ascension of human rights discourses sought to revive faith in a postwar new world order by ascribing rights to the human as such. Described by legal historian Samuel Moyn as the "last utopia," the gradual emergence of human rights in the postwar aftermath of catastrophe aspired legally to save the sovereign individual and its rights to self-determination.[11] Eventually adopted as the universal engine of social progress and justice—in the words of international law scholar Costas Douzinas, "by left and right, the north and the south, the state and the pulpit, the minister and the rebel"—reparations and human rights must be approached as a concerted, albeit inadequate, attempt to come to terms with the unfathomable legacies of human violence and erasure under colonial modernity.[12]

Chapter 2 transports us from postwar Europe to Cold War Asia, from the International Military Tribunals (IMT) in Nuremberg to the International Military Tribunals of the Far East (IMTFE) in Tokyo, and from the Transatlantic to the Transpacific. It brings together legal proceedings, reportage, and literature on war and violence in the Transpacific to explore the postwar ascension of reparations and human rights in regard to three interlocking historical events and their Cold War aftereffects: first, the atomic bombings of Hiroshima and

Nagasaki ending World War II; second, the internment of Japanese Americans by the US government during that war; and third, contemporary legal claims by "comfort women," young women and girls conscripted into sexual slavery by the Japanese Imperial Army.[13]

Unlike the Holocaust in Europe, reparations for the atomic bombings in Japan remain unthinkable to this day, and contemporary legal suits for apology and redress brought by surviving comfort women against the Japanese government have also been largely unsuccessful. In contrast, Japanese American claims for reparation against the US government have been comparatively effective, culminating in the Civil Liberties Act of 1988, whereby the US government apologized for internment and granted $20,000 to every surviving internee.[14] These three interconnected events—as well as the law's profoundly differential responses to their traumatic legacies—implicate overlapping histories of European colonialism, Japanese imperialism, and US empire in the Transpacific. In the same breath, they confound our assumptions about—indeed, our ability to distinguish between—victims and perpetrators. Rethinking the transnational possibilities and limits of reparations and the human after genocide, I bring the problem of racial reparations and the human into the irradiated space of the Transpacific. As numerous commentators have emphasized, the Cold War in Asia was anything but cold. To borrow a compelling concept from Asian American feminist theorist Mimi Thi Nguyen, the "gift of freedom" paradoxically bestowed upon this region and its stateless refugees a series of unending US-led military interventions, partitions, and wars connected to the rise of the American Century, European decolonization, and movements for postcolonial self-determination across the Third World.[15]

Chapter 2 continues the exploration of the quandaries of reparation by examining how psychoanalytic paradigms of trauma come to shape and to diverge from legal designations of victims and perpetrators deserving and undeserving of repair as well as juridical determinations of innocence and guilt. An analysis of European genocide, the IMT, and what the feminist historian Joan Wallach Scott describes as "the judgment of history" opens a new perspective on the social and psychic consequences arising from those foundational events. As Scott contends, that judgment configured the Holocaust as a singular incident, memorialized it as the epitome of evil, and mobilized Jewish claims to eternal victimhood.[16] To think the impossibility of the judgment of history in the space of the Transpacific, I turn to John Hersey's August 1946 *New Yorker* essay "Hiroshima," detailing the aftermath of the atomic bombing through the eyes of six surviving inhabitants. I also draw on the literary oeuvres of authors Kazuo

Ishiguro and Chang-rae Lee to connect nuclear devastation to internment and the comfort women system as well as to the conflicting postwar histories of civil rights in the United States and human rights on the world stage.

In contrast to the unanimous verdicts of guilt at Nuremberg, the dissension that marked the contentious judicial proceedings at the tribunals in Tokyo underscores how psychoanalytic approaches to the history of the (traumatized) subject supplement the subject of (Cold War) history still in search of a historical consensus. Here the psychic comes to subtend the legal. As psychic paradigms of trauma and guilt are mobilized to produce the proper juridical subject of human rights violation and crimes against humanity—the paradigmatic Holocaust victim—they simultaneously work to shore up the sovereignty of the injured human being as well as that of the victorious nation-state in the name of justice and repair. This shoring up of state and individual self-determination should not, however, be confused with an ethics of recognition.

There are undoubtedly situations where legal judgments must be made and punishments allotted. However, determinations of guilt and innocence, and the naming of victims and perpetrators, are not, as the philosopher Judith Butler observes, "the same as social recognition. In fact, recognition sometimes obligates us to suspend judgment in order to apprehend the other"—indeed, to apprehend the humanity of the other in order to apprehend a humanity *in* the other.[17] Ultimately, I argue, the suspension of historical judgment in Cold War Asia demands that we begin to theorize reparation and the human outside paradigms of sovereignty altogether. We must do so not for the sake of the victorious nation-state or its privileged citizen-subject but, rather, on behalf of those rendered inhuman by their fraught political legacies and loaded psychic dynamics.

Chapter 3, "Beyond Sovereignty: Absolute Apology/Absolute Forgiveness," offers one such theorization. It returns us to the Transatlantic by bringing together the aftereffects of nuclear fallout and environmental disaster in the Transpacific with an earlier scene of colonial violence in the Americas. As such, the book ends where it begins: with Indigenous dispossession in the New World. Here I turn to an account of uranium mining and nuclear arms production—in particular, the creation of Little Boy, the atomic bomb detonated by the US military over Hiroshima on August 6, 1945. While the long-lasting physical effects of atomic arms production, detonation, and testing become increasingly evident today, the unpredictable social effects of nuclear fallout remain unknowable. The toxicity of the Atomic Age devastates communities in radically uneven ways, even as it forms unanticipated social bonds and unpredictable alliances through the unwilled address of injured and unknown others.

Much of the world's uranium supply is mined from Indigenous lands. The uranium for Little Boy and the larger Manhattan Project, too, came in part from territories of the Sahtu Dene, an Indigenous people on the shores of Great Bear Lake in northwestern Canada. Most of the Sahtu Dene men who helped mine and transport the ore died of cancer. Their families all suffer from exorbitant rates of cancer. Like Oedipus, the Sahtu Dene were ignorant at the time of their actions—of the intended purpose of the ore they helped to extract or of its final destination. Nonetheless, they felt implicated once they learned many decades later of their connection to Hiroshima's fate. In response to the disaster, the Sahtu Dene sent a delegation to Hiroshima to apologize.

The Sahtu Dene's extraordinary response to the atomic bombing brings together a long history of Indigenous dispossession in the Americas with recent militarism and violence in the Cold War Transpacific. It highlights global framings of colonialism and race that bring together Asian, Asian American, and Indigenous communities in unexpected ways. In the same breath, their performative act raises the question of what it means to take responsibility for a historical catastrophe for which you are not quite responsible. Here I extend Jacques Derrida's notion of "absolute forgiveness" to create a corollary concept of what I call "absolute apology." If, as Derrida contends, absolute forgiveness demands forgiving something that is unforgivable, absolute apology involves apologizing for something for which you are not directly responsible.

Absolute apology thus offers an alternative model of reparations and the human eccentric to structures of state sovereignty, with its political calculations of cause and effect, its nominations of perpetrators and victims as figures for punishment and redress, its writing of an authorized history of violence into the past in the name of justice. To the contrary, the Sahtu Dene's actions illustrate a model of apology adjudicated not by the sovereign authority of the settler state and its sanctioned citizen-subjects but by those considered nonsovereign bad objects—the dispossessed, the exiled, the refugees, the stateless. It therefore situates the problem of repair beyond legal frameworks demanding the nomination of one victim and one perpetrator, assuming the clear innocence of the former and the incontestable guilt of the latter. It not only complicates ideas about repair and responsibility but also offers an alternative to them.

Separated from their land by colonial settlement, targets of enormous state violence and neglect, and victims of unfathomable environmental disaster, the Sahtu Dene nonetheless voluntarily adopt the mantle of a perpetrator, or what Jewish studies scholar Michael Rothberg describes in a different historical context as an "implicated subject."[18] They denaturalize the conventional legal boundaries separating victims from perpetrators in a long history of liberal reason and take responsibility for pain and misery that claim them as much as anyone else. Their actions thus highlight a nonsovereign model of repair, an ethics of living together not in resolution with the violence of the past but, rather, in continuous relation to its enduring traumatic legacies—to their permutating consequences in the present and their unforeseen effects in the future.

As they cleave to the victims of the atomic bombings in Japan, the Sahtu Dene assent to a notion of history, to borrow a concept from literary critic Cathy Caruth, as being implicated in each other's traumas.[19] Absolute apology thus reconsiders the universal aspirations of the human and human rights from the perspective of complicity and responsibility rather than the position of moral certitude and blamelessness. Along with Wynter, this book thus strives for "a redescription of the human outside the terms of our present descriptive statement of the human, Man, and its overrepresentation," without which we will never "unsettle [the] coloniality of power."[20] *Reparations and the Human* attempts to provide not only a description but also a redescription of the inherited political and psychic legacies of the human—of repair, trauma, and sovereignty—inside as well as outside its Western legacies and assumptions. Ultimately, the book underscores the fact that "we have never been human," to revise a concept from French philosopher Bruno Latour, suggesting that contemporary calls for the posthuman may be politically premature.[21]

I end this introduction by noting that there is a robust body of contemporary scholarship interrogating the prospects and liabilities of human rights and the redemptive power of the law after colonialism, genocide, enslavement, and occupation. Recent decades have witnessed various colonial as well as postcolonial governments across the globe addressing violations of the human and human rights as they attempt to confront histories of state-sponsored violence and terror, and as they seek to transition from authoritarian to more democratic modes of governance. Such violations, it is crucial to emphasize, are not restricted to any particular state form—capitalist or socialist, liberal or authoritarian, right-wing or left-wing, Western or Third World—in the Global North or Global South.

Notably, since their rapid ascension in the mid-1970s, discourses of universal human rights have sought to bind violations of the human specifically to authoritarian left-wing regimes. Human rights appeared on the scene precisely when they could be imputed to socialist rather than capitalist states, and when totalitarianism, as political theorist William Pietz has argued, could be detached from its origins in European fascism (Germany) and associated with "Orientalized" states (Russia) and left-wing governments in Asia and Latin America.[22] This shift has significant implications for the dream of decolonization, the will to racial justice, the hope for economic equality, and the politics of reparation.

Today discourses of universal human right function, in part, as a mechanism for the disciplining of postcolonial nation-states striving for self-determination through the legal judgments of an international community of former colonizers. At the same time, contemporary models of reconciliation and forgiveness attached to national apologies, truth commissions, and monetary restitution have yet to rectify the unfinished business of colonialism. A growing body of writings by scholars such as Stephen Hopgood, Nicola Perugini and Neve Gordon, and Randall Williams explores how discourses of human rights have been hijacked for all sorts of political ends.[23] In contrast, the topic of reparations has received comparatively less critical attention and, in contemporary scholarship and politics, tends to be either univocally lauded by progressives as an uncontested good or else skeptically dismissed by liberals and conservatives alike as a political impossibility. *Reparations and the Human* intervenes in these debates by examining the complex entanglements between political rights and economic redistribution embedded in colonial modernity and liberal social contract theory devoted to guaranteeing life, liberty, and property for sovereign European man. At the same time, my book explores their ongoing aftereffects in the Cold War Transpacific as a consequential limit case for the politics of repair and redress.

The history of reparations is as politically compromised as it is psychically fraught, despite the concept's present-day associations with justice. In short, we should neither conflate nor confuse a history of reparation in either political or psychoanalytic theory with the ethics of repair. Let us think back to the colonial history of reparations as political concept: among the three great eighteenth-century transatlantic revolutions—the French (1789–99), American (1775–83), and Haitian (1793–1803) Revolutions—the third stands out as exemplary in the history of the rights of man insofar as it was the most successful slave revolt in the world. Unlike their American and French counterparts, enslaved Haitians liberated themselves into freedom from their French colonial masters. Yet, as the Caribbean anthropologist Michel Rolph-Trouillot observes, the Haitian Revolution was forgotten even as it occurred, and in its volatile aftermath, Haiti had to pay France onerous reparations for the loss of its colonial possessions in Saint-Domingue.[24]

Today the long-lasting consequences of these economic concessions endure. Haiti continues to be the most impoverished nation in the Western Hemisphere, suffering from widespread poverty, disease, political instability, and lack of adequate infrastructure, tragically exposed each and every time an earthquake strikes the region. Similarly, after the US Civil War, the US government directed reparations to enslavers for the loss of their human property rather than to enslaved people who had been human property. Moving forward into the postwar period, although Germany paid significant reparations to Jews and to the state of Israel, the country has never paid reparations for the genocide of the Herero in Namibia. Similarly, Japan has yet to reckon with its own colonial histories in Korea and other parts of East and Southeast Asia, a history I will examine in chapter 2 of this book. Such omissions mark a national amnesia regarding crimes against humanity committed against the colonized—a pattern of nonrepair and nonrecognition rooted in New World discovery, continuing through Cold War Asia, and persisting into our postcolonial present.

What would it mean, then, to investigate reparations not as a solution to, but as a source of, post/colonial violence? What would it mean to analyze the concept as a key term by which aggression and violation is systematically directed and channeled according to a post/colonial logic of human and inhuman, civilization and savagery, blamelessness and culpability, and capitalist and socialist modernity across different times and geopolitical spaces? What would it mean to examine the human not as the basis of, but instead as a problem for, human rights and ethics? If there is nothing essential to being human beyond the fact that we coexist, remain vulnerable, and are governed in relation to one another by power, then how might we rethink reparations in terms of the political and psychic architecture establishing the modern, sovereign nation-state and its privileged citizen-subject? When does reparation become punitive and oriented toward revenge and punishment rather than restoring a social order of livable relationality? If the history of reparations and the human is profoundly compromised in both political and psychoanalytic theory, then how might we redefine the concept of repair in the service of emancipation and antisubordination? How might we approach the concept less as a noun than as a verb—that is, less as a settled event than as an unfinished process, an ongoing demand for justice for those erased from humanity? How do we come to know injury and harm? How are we to judge and redress them? In the final analysis, historical and conceptual dilemmas of reparations and the human index the problem of writing a consensual history of violence—of determining who is affirmed and who is forgotten in the aftermath of catastrophe.

In their attempts to repair past wrongs and injuries, nation-states engage in legal forms of apology as well as economic modes of redress meant to write a history of violence into the definitive past. However, a calculus of moral sentiment attached to national apology or material symbolism linked to economic payments meant to manage the suffering of others remains inadequate to the drive for justice. Psychic temporalities of trauma, loss, and suffering do not operate through the logics of juridical and financial accountability, nor can they be resolved by the sovereign authority of the nation-state and its self-determined liberal subject. Ultimately, if the past is irreparable, then a reconceptualized notion of reparations must concern itself in the present with creating the conditions of possibility for just futures and for new relationalities to be sustained. As renewed demands for reparations come to occupy center stage in debates concerning unresolved legacies of dispossession and transatlantic slavery, a critical reassessment of reparations from a comparative perspective is long overdue. This book is my attempt to provide one from the vantage point of Cold War Asia by investigating reparations and the human not as a moral response to but, rather, as a form of continued state violence.

1

Beyond Repair

Political and Psychic Genealogies in Locke and Klein

But we permitted ourselves to have other hopes. We had expected the great world-dominating nations of white race upon whom the leadership of the human species has fallen, who were known to have world-wide interests as their concern, to whose creative powers were due not only our technical advances towards the control of nature but the artistic and scientific standards of civilization—we had expected these peoples to succeed in discovering another way of settling misunderstandings and conflicts of interest. . . .

Then the war in which we had refused to believe broke out, and it brought—disillusionment. —SIGMUND FREUD, *Thoughts for the Times on War and Death* (1915)

In 1920, Sigmund Freud outlined his famous theory of the death drive in *Beyond the Pleasure Principle*.[1] This short but powerful volume was composed in response to the pattern of unremitting trauma exhibited by soldiers returning from the battlefront during World War I. Similar to Freud's civilian patients, these military combatants suffered from "reminiscences." Their reminiscences, however, could be traced not to a primordial or pleasurable object of desire but, unexpectedly, to these soldiers' recent participation in grisly violence on the battlefield. Witnessing the involuntary reliving of their painful experiences—what came to be diagnosed clinically as "shell shock" in his time and what has evolved into "posttraumatic stress disorder" in ours—Freud wrote *Beyond the Pleasure Principle* to understand better the human drive toward death. A decisive turning point in his metapsychology, Freud's slim volume raised disturbing questions about a repetition compulsion outside of conscious control, one overriding, irreconcilable to, and indeed *beyond* the pleasure principle: compulsive and unacknowledged behavior that oriented its subjects not toward life but toward death.[2]

The initially patriotic Freud would soon come to lament World War I—during which two of his sons and a son-in-law volunteered in the Austrian army to fight on behalf of Germany—as so "cruel," "embittered," and "implacable" that it "tramples in blind fury on all that comes in its way, as though there were to be no future and no peace among men after it is over."[3] The war brought "disillusionment," with Freud and many other leading thinkers of the day losing faith in the project of European civilization and progress associated with the "great world-dominating nations of white race upon whom the leadership of the human species has fallen."[4] This war was a global conflagration in which techniques of violence honed in outlying colonial regions returned to haunt Europe. It inaugurated the concept of "weapons of mass destruction," and it exercised these new instruments of human annihilation liberally on soldiers as well as civilians in Europe. World War I created massive populations of the stateless and refugees; it exemplified the contingency of natural rights and anticipated Hannah Arendt's trenchant observation, to return to the opening citation of my book, that "the world found nothing sacred in the abstract nakedness of being human."[5] Ultimately, the war helped pave the way for another repetition compulsion of unfathomable horror less than two decades later: genocide on the continent under fascist rule and the even more total destruction of World War II following the rise of the Third Reich.[6]

Scholarship in postcolonial studies and new European history links the Holocaust and its intensified violence to a string of colonial genocides in Africa, Asia, and the Americas as part of a "racial century" (1850–1950) leading up to the "Final Solution."[7] Critics of French colonialism, including Aimé Césaire, Frantz Fanon, and Octave Mannoni, and leading US thinkers on Black internationalism such as W. E. B. Du Bois, narrate the rise of European totalitarianism as a kind of colonial project turned inward.[8] In his preface to Fanon's *The Wretched of the Earth*, Jean-Paul Sartre describes colonialism as a "boomerang" that returned to plague Europe in the form of fascism.[9] In this regard, the ascension of the National Socialist German Workers' Party during the interwar years—typically characterized as the most extreme example of such total violence—was only the climax of a recursive social process that, rather than dividing, marked and bound together all European colonial powers to varying degrees.

From this angle, it would not be inaccurate to describe the genealogy of Freud's death drive as embedded in a haunting, though unacknowledged, colonial history—one with which the field of psychoanalytic theory and practice has fully yet to reckon. Importantly, Césaire, Fanon, and Mannoni delineated not just the material but also the psychic ravages colonialism exacted on Native subjects and their subjectivities. Similarly, Freud's theory of the death drive might be considered the psychoanalytic corollary of Sartre's "boomerang," a detailed exegesis of the damaging effects that war, violence, and colonialism had on the European psyche. In short, Freud's death drive marked the *psychic* dissolution of the liberal human subject and its storied history of consciousness.

In slightly different terms, the death drive links the dissipation of European liberalism and its elusive dream of enlightenment with a traumatic history of colonialism erupting in, as well as symptomized by, the fractured psyches of its shell-shocked combatants. Notably, from this perspective, trauma is configured as a kind of naturalized property of the liberal human subject. Attached to Freud's injured male war veteran, it carves out a privileged space of exhausted and victimized humanity with significant implications for repairing the injuries of violated human beings in Europe and elsewhere.

The uneven distribution of trauma—and of violence, victimhood, and repair—marks an implicit albeit understudied dynamic structuring various political and psychic assumptions regarding the differential production of the human being and its rights to reparation. Indeed, I argue throughout this book, the right to reparation comes retroactively to produce the figure of the human being itself. This dynamic emerges from the advent of New World discovery and the rise of the Enlightenment to continue through the Enlightenment's demise in holocaust and nuclear devastation. The unremitting psychic negativity binding a history of colonial violence with the unacknowledged arrogation of traumatized humanity defied Freud's comprehension of the death drive—its psychic and political aporias. It continues to defy ours.

Beyond the Pleasure Principle marks a significant epistemological break in Freud's thinking. It moves us beyond purely sexual etiologies of trauma and neuroses by forcing us to rethink not only the psychic but also the social parameters of European civilization and progress in the face of an unrelenting history of global war and violence binding Europe together with Africa, Asia, and the Americas under colonial modernity. From his treatment of female hysterics confined to the sequestered chambers of the bourgeois home, which inaugurated his concept of the unconscious, to his ministering to wounded male combatants returning from the ravaged battlefront, Freud pushed the boundaries and limits of psychoanalysis's ability to address pain and repair misery. The death drive marks the emergence of this impasse in Freud's thinking, but it also symptomizes a much wider social—as well as much longer psychic—history troubling the problem of reparations and the human globally.

I begin this chapter by investigating the production of reparations and the human in the context of a colonial history that both conjoins and separates Europe and the Americas. Here I bring together John Locke (1632–1704) and Melanie Klein (1882–1960), two key thinkers in political theory and psychoanalysis who are rarely discussed in relation to one another, by considering their formative writings on reparation in tandem. Together, their theories on reparation delineate the political evolution of the rights of man and its attendant psychic history of consciousness. As we shall see, reparation in Locke and Klein—and, indeed, the development of liberal human subjectivity itself—is predicated on the systematic management and displacement of colonial aggression and injury. The emergence of Enlightenment thought and its hallowed principles of human freedom—of rights to life, liberty, and property—cannot be dissociated from the project of settler colonialism and its protocols of repair.

The first section of this chapter, "The Political: Locke's Two Bodies," begins with an analysis of Locke's 1690 *Two Treatises of Government*. Here I consider how Locke's writings on reparation establishing the political foundations of the rights of man framing the eighteenth-century Enlightenment project are grounded in a bloody history of Indigenous dispossession in the Americas. The *Two Treatises* are traditionally interpreted as an outline of a universal theory of natural rights, the emergence of the social contract, and the institution of civil society through collective consent of the governed. In contrast, I focus on the ways that Locke's account of the evolution of property rights and concomitant claims to life and liberty in Europe and the Americas are radically and unevenly distributed across the Transatlantic as a function of institutionalized violence and entitlements to repair.

The second section of this chapter, "The Psychic: Klein's Colonial Object Relations," brings us forward from Locke's early Americas via Freud to the twentieth century and to Klein, who began to develop her own theories of idealization, aggression, destruction, and repair during the interwar years as the dream of enlightenment dissolved into ashes. The contemporary turn to affect in queer theory pioneered by queer studies scholars such as Eve Kosofsky Sedgwick and Lauren Berlant emphasizes the redemptive quality of love and "cruel optimism" in the reparative process.[10] In my investigation, however, I explore this dynamic of love and repair by offering a very different reading of affect in Klein, a process I describe as "colonial object relations." Attention to colonial object relations outlined in Klein's classic 1937 treatise on "Love, Guilt and Reparation" symptomizes the ways affect is unevenly distributed in the history of liberal empire and reason. It reveals how love and hate are affectively policed to create a field of good and bad objects, as well as liberal and Indigenous subjects, regulated by a colonial morality that is not the cause but, rather, the effect of processes of reparation. It underscores how this colonial morality comes to define liberal bad faith and white guilt. Together Locke and Klein configure the liberal human subject as a traumatized victim deserving of repair while constituting the Indigenous other as an aggressive perpetrator deserving no consideration.

What, then, are the colonial dimensions of object relations or, more generally, the affective contours of liberal reason?

The Political: Locke's Two Bodies

> A Planter in the *West Indies* has more, and might, if he pleased (who doubts?) Muster them [sons, friends, soldiers, slaves] up and lead them out against the *Indians*, to seek Reparation upon any Injury received from them. —JOHN LOCKE, *Two Treatises of Government*

Who is deserving of repair? How do law and politics adjudicate this problem, and who do they protect from harm—from past injuries as well as injuries yet to come? Let us turn to Locke's conception of reparations in *Two Treatises of Government*, a cornerstone of liberal social contract theory, to examine these questions.

Recent years have seen excellent scholarship in political theory exploring the intertwined relationship between liberalism and colonialism in the context of New World conquest and settlement.[11] David Armitage, Peter Laslett, Uday Mehta, and James Tully, for instance, have all written eloquently on the relationship of property to Native dispossession in Locke's oeuvre.[12] They raise the question of how a history of colonialism might be considered constitutive rather than peripheral to the rise of liberalism and to the *Déclaration des droits de l'homme et du citoyen*, a key document of democratic revolution with which postwar discourses of reparations and human rights are often associated.[13] These debates draw specific attention to the exclusionary bases of liberal theory in the face of its putative claims to abstract equality and freedom—that the natural rights of man, as the *Déclaration* puts it, "are held to be universal and valid in all times and places."

Collectively, these political theorists ask whether the "historical Locke" can ultimately be separated from the "theoretical Locke." The historical Locke served in various colonial administrations throughout his career, including as Secretary of the Board of Trade and Plantations and Secretary to the Lords and Proprietors of the Carolinas (under his patron Anthony Ashley Cooper, later the First Earl of Shaftesbury). Locke profited richly from various colonial enterprises in the Americas, especially from his investments in the slave trade through the Royal African Company and his standing as a landlord. Through his political and philosophical writings the theoretical Locke, along with Thomas Hobbes (1588–1679) and Jean-Jacques Rousseau (1712–78), helped to establish eighteenth-century Enlightenment precepts for liberalism and the limitation of state power: the social contract, the rights of man, and consent of the governed.[14] The commingling of these two aspects of the historical and the theoretical Locke, Armitage notes, provided him with

> a more thorough understanding of his country's commerce and colonies than that possessed by any canonical figure in the history of political thought before Edmund Burke. No such figure played as prominent a role in the institutional history of European colonialism before James Mill and John Stuart Mill joined the administration of the East India Company. Moreover, no major political theorist before the nineteenth century so actively applied theory to colonial practice as Locke did by virtue of his involvement with writing the *Fundamental Constitutions* of the Carolina colony. For all these reasons, Locke's colonial interests have been taken to indicate that "the liberal involvement with the British Empire is broadly coeval with liberalism itself."[15]

As Armitage notes, references to America and its Native inhabitants appear throughout the second of Locke's *Two Treatises*. They populate seven of the eighteen chapters of this volume, with more than half of these references clustered around chapter 5, "Of Property."

For Locke, the emergence, evolution, and protection of property rights is key to the establishment of civil society under liberalism based not on the divine right of kings but on the mutual consent of the governed. In *Two Treatises*, the state of nature is a state of reason. It is precisely through individual reason that every man in a state of nature first claims self-possession and "*Property* in his own *Person*."[16] He subsequently transforms this property in his own person into "the *Labour* of his Body, and the *Work* of his Hands," in the process removing objects and things from "out of the State that Nature hath provided."[17] Mixing it with his labor, which "*puts the difference of value* on every thing," he thereby transforms them into his property.[18] His most famous theoretical contribution to the coterminous establishment of liberalism and capitalism, Locke's conceptions of property delineate the path by which, as Laslett observes, "men can proceed from the abstract world of liberty and equality based on their relationship with God and natural law, to the concrete world of political liberty guaranteed by political arrangements."[19]

America provides the historical backdrop for this theoretical transformation of natural law into political reality as well as economic value. In Locke's famous estimation of this evolution of civil society and property, "in the beginning all the World was *America*."[20] The New World serves as the privileged site for the unfolding of self-possession through enclosure of the commons inaugurating Locke's ideas on individual self-realization through his labor theory of value and the development of the social contract. Importantly, as the land of plenty in a state of nature, the commons in America may be appropriated without the consent of its Native inhabitants. Locke positions the colonial settler as the idealized archetype for this transformation of untrammeled land into private property through his industrious labor: "God gave the World to Men in Common; but since he gave it them for their benefit, and the greatest Conveniences of Life they were capable to draw from it, it cannot be supposed he meant it should always remain common and uncultivated. He gave it to the use for the Industrious and Rational, (and *Labour* was to be *his Title* to it;) not the Fancy or Covetousness of the Quarrelsome and Contentious."[21]

Given the prominence of the New World in Locke's historical account of the development of property and civil society, not to mention his direct involvement in the transatlantic slave trade, it is hardly surprising that critical discussions of the relationship between liberalism and colonialism have focused primarily on this convergence—one, I would argue, that functions as a philosophical base for the defense of racial capitalism. The individual's cultivation of property, along with the development of property rights in the context of the New World, demanded the continuous invocation, management, and exclusion of the figure of the "Indian in the Woods of America." While arguments concerning Indigenous dispossession are well rehearsed in liberal political theory and postcolonial studies, they nonetheless bear repeating.

Colonial arrogation of Indigenous land, as various critics have noted, is facilitated precisely through Locke's refusal to recognize Native forms of labor, as well as Indigenous techniques of land management, cultivation, and conservation, as bearing property rights or as reflecting commensurate forms of political organization and governance for Native populations. Tully, for instance, situates this refusal in a colonial discourse of development. He observes, "Locke defines property in such a way that Amerindian customary land use is not a legitimate type of property. Rather, it is construed as individual labour-based possession and assimilated to an earlier stage of European development in a state of nature, and thus not on equal footing with European property. Amerindian political formations and property are thereby subjected to the sovereignty of European concepts of politics and property."[22] Subordinated to European political ideas of sovereignty and property, Indigenous forms of social organization and labor are consigned to what the postcolonial historian Dipesh Chakrabarty describes as "the waiting room of history."[23] European sovereignty comes to overwrite Indigenous relations and claims to their own lands by relegating them to an antiquated past and by subordinating them to the labor of the European settler and possessive individual. Simultaneously, it places Native populations in a state of exception to European law, preconditioning rather than describing all human beings and their fundamental rights under colonialism. In Locke's pithy assessment, a "king in America is worse off than a day labourer in England."[24]

The political exclusion of the Native—not to mention slaves, women, and those without sufficient reason to exercise suffrage and self-possessive individualism—over the past three and half centuries, observes Mehta, "must be allowed to embarrass the universalistic claims of liberalism."[25] Yet in all these discussions, very little has been said about how the arrogation of Indigenous lands required not only the enclosure of the commons in a state of nature but also the systematic dispossession and death of Indigenous populations through a naturalized process of unwarranted aggression and repair. This process required that Locke constitute reparations as a key term to redress the injuries suffered by defensive colonial settlers, the "industrious and rational," at the hands of aggressive Natives, the "quarrelsome and contentious." For Locke, reparations represented the political right to exterminate the Natives and appropriate their lands. To situate this dynamic in a comparative historical perspective, unlike "tractable" Indigenous populations in Mexico, Central, and South America, whose religious conversion and labor underwrote the *encomienda* system imposed under Spanish colonial rule, belligerent Natives in North America impeded British settler colonialism and its transatlantic slave-based labor system.[26] Their incessant invocation as bellicose and irrational thus served to justify their endless removal and elimination.

I explore how the politics of reparations in Locke became a fundamental concept for rationalizing intertwined processes of dispossession and genocide of Native populations in North America. Here I focus on Locke's theory of reparations as it unfolds in the less-analyzed chapter 16 of the *Second Treatise*, in which Locke juxtaposes his discussion of the rights to repair with his more familiar accounts of property outlined in chapter 5. In so doing, Locke might be said to rework the medieval idea of the "King's two bodies," subtending the political theology of monarchical divine right into his account of reparations and social contract theory. However, unlike the king's corporeal and eternal bodies, reparation in Locke assumes "two geographical bodies"—two different meanings—across Europe and the Americas.[27]

When the duality of reparation in Locke travels across the Transatlantic, its function shifts from protecting the liberal human subject from the capricious violence of royal despotism in Old World to authorizing limitless war against Indigenous populations in the New World. In other words, through its transatlantic displacement between Europe and the Americas, reparation affirms the prerogatives of the liberal human subject in the face of royal excesses while occluding a history of unchecked violence against the Natives in the face of conquest. Indeed, by conceptualizing reparations as justification for colonial defense in the face of Indigenous assault, Locke produces the European settler in the Americas as a traumatized victim of Native aggression—as the paradigmatic injured and aggrieved subject of colonial modernity.

Before turning to the *Two Treatises*, I offer a brief history of the concept of reparations. Classical definitions can be traced to the principle of *lex talionis* (law of retaliation) or, in its later English translation and transformations, the "law of talion" (retaliation authorized by law). In the Old Testament, the notion of "an eye for an eye" describes an injured party receiving the same in kind as compensation. Similar to the Abrahamic tradition, the Code of Hammurabi (1780 BCE) imposed limits on retribution after harm. As long as perpetrator and victim were of equal social standing, Babylonian law stipulated that the punishment could not exceed the crime. Such legal standards of reparation are conventionally interpreted as restricting excessive or arbitrary retaliation at the hands of an avenging party, a principle of exact and exacting reciprocity.

Under Anglo-Saxon penal law that emerged in the sixth century, the idea of reparations acquired its economic inflections, expanding to include monetary compensation over direct retribution, with a value affixed to every person or piece of property. These practices of restitution as redress evolved from the principle of *wergild* (literally, "body payment") in Salic law, the most influential branch of Germanic laws (*leges barbarorum*), one translating noneconomic losses into monetary equivalents through actuarial tabulations calculating everything from injury to death. Over time, these practices formed the modern legal foundation for torts in common law—the field of civil legal liability concerning a wrongful act or an infringement of a right outside of contractual agreement.[28] Unlike contract law, whereby obligations are determined in mutual agreement between two independent private parties, reparations for harm in tort law, similar to criminal law, are imposed and regulated by the state. This connection between torts and criminal law forms the modern-day legal foundation for the postwar emergence of reparations and human rights in international law under neologisms of *illegal war*, *war crimes*, and *crimes against humanity*.

On the one hand, when considered as restricting excessive punishment and as exacting fair compensation for injury and loss, reparations as a tool of justice situate the law as a disinterested mediator between two independent parties within one shared juridical system. In this regard, it constitutes a mechanism internal to the law for establishing moral accountability between abstract and equal subjects of a sovereign entity. This history of accountability and abstract equality emerges under European modernity for the self-determined liberal individual. On the other hand, insofar as the concept also applies to interstate conflicts between two opposing powers, the term *reparations* also describes the process whereby a victorious state compelled a defeated one to pay for the costs of war. This convention dates back at least to the first Punic War (264–41 BCE), when Rome imposed monetary payments on a defeated Carthage.[29]

Here notions of restraint on compensation, punishment, and vengeance—along with attendant problems of justice—reach a conceptual limit. Originally, reparations were often considered little more than the spoils of war. As Thucydides observes in the "Melian Dialogue," his famous commentary on the Peloponnesian War, "The strong do what they have the power to do and the weak accept what they have to accept."[30] We might regard this history of victor's justice as framing the problem of reparations and the human under colonial modernity. As I suggested in the introduction, a logic of colonial primacy continued to govern not only in the aftermath of the Haitian Revolution but also in the wake of the Holocaust and Hiroshima, as I will elaborate in chapter 2. As such, the postwar rise of reparations and human rights as a concerted moral response to European catastrophe and the abstract nakedness of being human is a narrative of aborted beginnings, one tracing its troubled genealogy to the formative political writings of Locke.

The term *reparation* appears repeatedly across both of the *Two Treatises*, helping to establish Enlightenment concepts of repair and redress under European modernity. The majority of references to reparations occur in the *Second Treatise*, in chapter 2, "Of the State of Nature," and chapter 16, "Of Conquest." In the former, Locke outlines the principles of self-defense, punishment, and repair in a state of nature. In the latter, he amends traditional conceptions of reparations as the spoils of war in the context of European civil society by placing strict limits on the redistribution of property and rights to capital punishment after violent conflict.

Locke composed the *Two Treatises* (and published them anonymously) in the aftermath of the bloody revolutions and civil wars of seventeenth-century

England and in the wake of the restoration of the Stuart monarchy upon King Charles II's return from exile on the Continent in 1660. From a larger historical perspective, the *Two Treatises* are also responding to post-Reformation religious wars that wracked Europe until the signing of the Treaty of Westphalia in 1648, which I discuss further below. In what he delineates as the rules of just war, Locke restricts the prerogatives of the victor in the name of reparation. In chapter 16, "Of Conquest," he argues,

> For it is the brutal force the Aggressor has used, that gives his Adversary a right to take away his Life, and destroy him if he pleases, as a noxious Creature; but 'tis damage sustain'd that alone gives him Title to another Mans Goods: For though I may kill a Thief that sets on me in the Highway, yet I may not (which seems less) take away his Money and let him go; this would be Robbery on my side. His force, and the state of War he put himself in, made him forfeit his Life, but gave me no Title to his Goods. The *right* then *of Conquest extends only to the Lives* of those who joyn'd in the War, *not to their Estates*, but only in order to make reparation for the damages received, and the Charges of the War, and that too with reservation of the right of the innocent Wife and Children.[31]

The passage raises a number of curious paradoxes. Initially, Locke lays out a theory of reparation that limits in very specific ways the redistribution of property and life after violent conflict. More specifically, he constrains the prerogatives of the winner (whom he names the "Adversary") by protecting the rights of the loser (whom he labels the "Aggressor").

Here reparation extends only to those men who "joyn'd in the War, *not to their Estates*, but only in order to make reparation for the damages received, and the Charges of War, and that too with reservation of the right of the innocent Wife and Children." That is, even if reparations can be exacted for the "charges of war," and even if the Aggressor can sustain such costs, this redistribution of property is constrained further by the needs of "the innocent Wife and Children," who must not be harmed physically or materially in the aftermath of conflict and the restoration of peace. Indeed, reparation comes to guarantee their well-being and sustenance while trumping any legitimate claims the Adversary might have on the Aggressor's property.

The Adversary, in other words, has no right to ruin the Aggressor's dependents or to punish them for transgressions in which they did not directly participate. Even more, reparations cannot justify the Adversary's jurisdiction over the Aggressor's subjects and dependents. From this perspective, we might say that Locke diminishes the spoils of victory such that reparation emerges as the name for that political process by which the redistribution of property and life following conflict is pursued in the name of justice—as opposed to thievery—for both the Adversary and the Aggressor alike. Indeed, *reparations* here becomes the key term separating the state of war (thievery and exception) from the state of nature (reason and normalcy).

While proffering this definition of reparations as a limitation on the redistribution of property as well as deference to the life and liberty of "the innocent Wife and Children," Locke simultaneously introduces into his theory of repair a rather startling relationship between property and human life, one that becomes actualized, I suggest, only retrospectively in relation to the New World. The Aggressor, defined earlier in this passage by Locke as "thief" who sets upon his Adversary on the highway, may not forfeit his property, yet may need to give up his life: "For though I may kill a Thief that sets on me in the Highway, yet I may not (which seems less) take away his Money and let him go; this would be Robbery on my side. His force, and the state of War he put himself in, made him forfeit his Life, but gave me no Title to his Goods."

Here continuity between self-possession and industrious cultivation of the commons in a state of nature creating the foundation for Locke's theory of property and personhood in chapter 5 reaches a conceptual limit. The "brutal force the Aggressor has used" disrupts the continuity between personhood and property through the abuse of its fundamental principles, placing him in a state of exception. In mandating that the Aggressor forfeit his life but not his things, his property, Locke thus carves out two disparate spheres of life—of privileged and precarious life—separating Adversary from Aggressor.

In other words, Locke's theory of reparation implies that certain lives are less valuable than certain things. While the Adversary can kill a thief who attacks him on the highway, he cannot simply take away the thief's money and set him free, which Locke admits with a certain amount of (parenthetical) anxiety would "seem less" but would in fact constitute "robbery." Property thus emerges as an inviolable subject in and of itself; its sovereign sanctity and protection rise above that of particular lives. In effect, Locke's theory of reparations justifies the capital punishment of certain individuals and populations constituted as "Aggressors" and "thieves" with illegitimate relations to property. Thus, we begin to recognize how reparation in Locke marks the coming together of disavowed forms of Indigenous possession with genocide in the New World precisely by distinguishing between privileged and precarious life in a state of war.

Locke, of course, composed "On Conquest" in the historical aftermath of the English Civil War, the Restoration, and the Glorious Revolution. From this perspective, in seeking to restrict the royal prerogatives of the Stuart monarchy to punish those who opposed the regency, to seize their property, and to destroy their lives, Locke's theory of conquest seeks to secure the welfare of the European citizen-subject and progeny. The "Father, by his miscarriages and violence, can forfeit his own life, but involves not his children in his guilt or destruction."[32] The Adversary has no right of dominion over the lives or property of the Aggressor's descendants. What must be emphasized in my analysis, however, is how the late seventeenth- and eighteenth-century emergence and protection of sovereign liberal rights is subtended by a realm of devalued and precarious life, one in which the redistribution of property and the meting out of punishment remains *limitless*. I am, of course, referring not to conquest in general but to *colonial* conquest in particular.

How is the affirmation of an "internal" history of reparations and the protection of the rights of (hu)man and citizen in Europe radically transformed by the forgetting of this "external" history of reparation as unrestrained aggression against Indigenous populations in the New World? How does Locke's theory of reparation come paradoxically to assume two geographical bodies—to constrain as well as to authorize violence in its transatlantic ambit? How does the figure of the Adversary—whom I would describe in modern terminology as the "traumatized victim"—work to secure this dynamic? Let me turn for a moment to conventional accounts of the origins of modern international law and, in particular, European state sovereignty as the right to wage war through the monopolization of violence.

Standard histories of the concept of European state sovereignty trace their birth to the Treaty of Westphalia in 1648, established during Locke's early adulthood. The treaty marked the end of post-Reformation religious wars in Europe. Legal historians see it as having terminated the eschatological temporality legislated by the Holy Roman Empire's universalizing aspirations. The treaty marked the initiation of European nations into a shared secular time frame in which religious self-determination defined the emergence of state sovereignty. That is, religious self-determination created formal equality among a European family of nations by secularizing both international relations and war as a matter of state rather than religious authority. Within this framework, we might consider how the social contract among self-possessive liberal individuals converges with a modern regime of international law among a European family of nations. In both instances, the secularization and arrogation of violence by the state occurs contractually on both the individual level (through the social contract) and the collective level (through a regime of modern international law and consensual recognition of state sovereignty).

Yet we might also consider the German jurist Carl Schmitt's claim that the "discovery" of the New World in 1492, rather than the Treaty of Westphalia, is a more appropriate historical moment to which the origins of state sovereignty, the monopolization of violence, and modern international law might be traced.[33] Deploying Schmitt's geographic and temporal framing of 1492 from an opposing perspective, the comparative legal theorist Teemu Ruskola observes that the narrative of the law of nations is "no longer one of increasing inclusion and equality *within* Europe. Rather, it becomes a story of the violent exclusion of others *outside* Europe, first on the basis of religious [the logic of Spanish and Portuguese colonialism], then cultural difference [the logic of English colonialism]."[34]

Commenting on the formative writings of the Spanish theologian Francisco de Vitoria, widely considered the father of international law (along with Hugo Grotius and Alberico Gentili), the international law scholar Antony Anghie summarizes this dynamic between European nation-states and colonized Natives in the Americas in a similar manner: "International law, such as it existed in Vitoria's time, did not *precede* and thereby effortlessly resolve the problem of Spanish-Indian relations; rather international law was created out of the unique issues generated between the Spanish and the Indians."[35] For Anghie, the emergence of modern international law introduces a number of pressing questions concerning the legitimation of violence as war became the "ultimate prerogative" of sovereign state and sovereign authority: "Who may wage war? When can war be waged? What limits must be observed in the waging of war? What constitutes a just war?"[36]

From the North American context, Locke's theory of reparations defining Anglo-Indian relations might be considered a response to these formative questions regarding the origins and concerns of international law. That is, reparations designate the political limits of legitimate violence and the politics of redress that can be pursued in the name of justice. If the rise of international law and the emergence of European sovereignty as the monopolization of both violence *and* reparation acquired its historical and theoretical consistency through the colonial encounter, we might describe reparation as constraining violence, limiting the redistribution of property, and protecting human life within Europe while externalizing and projecting this violence through the destruction of (in)human life and the seizing of Native lands outside Europe. As Locke asserts in the epigraph with which I begin this section, "A Planter in the *West Indies* has more, and might, if he pleased (who doubts?) Muster them [sons, friends, soldiers, and slaves] up and lead them out against the *Indians*, to seek

Reparation upon any Injury received from them."[37] The figure of the Indian in the woods of America is worthy of neither protection nor repair.

More specifically, reparation in Locke assumes two *cultural* bodies through the trope of civilization and savagery: the colonial settler with human reason, "industrious and rational," versus the barbaric Indian without human reason, "quarrelsome and contentious." Notably, the logic of Spanish and Anglo settler colonialism diverges. On the one hand, the Spanish Vitoria supplants divine law with natural law as the foundation of international law, one presuming universal reason among all peoples. Vitoria approaches Natives as either possessing reason (and souls) and therefore potentially part of a Catholic religious order or, alternatively, without reason (and barbarous), unwilling to convert, and therefore subject to holy war. While universal reason identifies the Indians of the New World as abstractly equal subjects in the *jus gentium* of the Spanish colonial order, Vitoria imagines them as rational subjects precisely either to discipline them into Spanish norms or to destroy them through "just war"—a form of legal inclusion precisely for exclusion.[38]

On the other hand, Locke's Anglo understandings of Native populations conceived the Indian as eccentric to civilized humanity and reason and therefore excludable from civil society. Throughout the *Second Treatise*, Locke presents us with a set of ghostly figures constituting the limits of the state of nature and its laws of reason. At times Locke calls these figures "brutes," but he also refers to them as "Indians," along with "Lunaticks, Ideots, Madmen, servants, children, slaves." The figure of the Indian in the woods of America exposes a sharp, internal limit to Locke's foundationalism, separating man from brute, human from inhuman.

According to Locke, when an individual violates human reason, he forfeits his natural rights and may be enslaved or killed. As Tully suggests, while "scholars who work on this part of Locke's theory assume that it refers to black slavery," it may also refer to Indian dispossession and death. Thus, Tully continues, while one strategy of Indigenous dispossession and extermination involved characterizing America as *terra nullius*—empty and vacant land—another "strategy was to downgrade the status of the aboriginal peoples to that of beasts or savages so no legal recognition was required. Often a royal grant would simply extend explorers and invaders the right 'to subdue, occupy and possesse' the inhabitants."[39] For Locke, as Laslett summarizes, "any man who behaves unreasonably is to that extent an animal, and may be treated as such."[40] Writing about the "persistent raciality" of the human/animal distinction overshadowing the transatlantic slave trade, Black studies scholar Zakiyyah Iman Jackson argues that the "repudiation of 'the animal' has historically been essential to producing classes of abject humans" subject to death.[41] Similarly, when the Indian is placed outside human reason, Indigenous populations can be treated as animals to be exterminated rather than as humans with rights to repair.

This geographic distribution of reparations in Locke creates two domains of valued (colonial) and precarious (Indigenous) life as a bifurcated operation of violence and repair. Even more, Locke's theory of reparations and the human fixes the Indian as "Aggressor" and the colonist as "Adversary" in his nomenclature—and as "perpetrator" and "victim" in my own. As an aggrieved and traumatized victim defending *his* property against Indian aggression, the colonial settler knows no bounds, literally.[42] For Locke, the colonizer never occupies the position of Aggressor-perpetrator; rather, he is aligned with the position of injured Adversary-victim deserving reparations. Equally so, in the *Second Treatise* there is no possibility that the Native might be the Adversary-victim defending his life and property. The historical or, perhaps more accurately, anthropological Locke has already assured us of this cultural division.

Through an insistent reversal and projection of this Adversary/Aggressor and victim/perpetrator dynamic, the colonial settlers defend their lives and property against savage Indians precisely by seizing their lands and destroying their lives as a noxious creatures, brutes, and animals. Reparation in America thus describes a retroactive process of Indigenous dispossession and death by marking a broken bond between personhood and property, a cleaving of human from inhuman, human from animal, colonial settler from Native Indian.[43] In other words, while the "Indian in the woods of *America*" is repeatedly invoked throughout chapter 5—indeed, helping to constitute Locke's general theory of property and property rights—by chapter 16 this figure has altogether vanished from the narrative of European progress and development of civil society and governance. Ultimately, this disappearance of the Native, along with the elevation of objects (property) over certain subjects (Indian in the woods), symptomizes the ways in which the Indigenous are invoked and instrumentalized—forgotten, erased, and buried—to create the ground for the emergence, affirmation, and figuration of the liberal human subject itself.[44]

The argument I trace above comes to serve as the theoretical foundation for Chief Justice John Marshall's landmark US Supreme Court decision *Cherokee Nation v. State of Georgia* (1831). Notably, in the spirit of Locke, Marshall refuses to acknowledge violence against the Cherokees as violence or to recognize their claims of injury as deserving of protection or repair: "If it be true that the Cherokee nation have rights, this is not the tribunal in which those rights are to be asserted. If it be true that wrongs have been inflicted, and that still greater are to be apprehended, this is not the tribunal which can redress the past or prevent the future."[45] Here the question of the "inhumanity" of the Indian takes a legal sleight of hand as the absence of a recognizable political community for the Cherokees and therefore a decided lack of standing before the law. After all, one cannot stand if one is buried.

More specifically, in describing the Cherokees as a "domestic dependent nation," Marshall's majority opinion creates a new constitutional category—a state of exception—that indexes Locke's history of reparations and the human in the Americas, while also foreclosing a future for a justice to come. The chief justice places the Cherokees as an "Indian tribe" in a legal abyss, in a state of pupilage in which they are neither a sovereign "foreign Nation" nor a domestic US state "among the several States" but, rather, a "domestic dependent nation." In other words, although they are signatories to legal treaties with the US government, Cherokees are neither sovereign citizen-subjects with political rights guaranteed by US federalism nor part of a sovereign European family of nations that can make claims against other states or engage in legitimate war. "The notion that indigenous peoples are *weaker than, wards, dependent,* and *limited* in power," Lenape scholar Joanne Barker observes, "has perpetuated dominant ideologies of race, culture, and identity" that constitute "a political strategy of the nation-state to erase the sovereign from the indigenous." This erasure, she argues, "is the racialization of the 'Indian.'"[46]

Marshall begins his majority opinion with a stark acknowledgment of a continuous history of injury and harm: "If courts were permitted to indulge their sympathies, a case better calculated to excite them can scarcely be imagined. A people once numerous, powerful, and truly independent, found by our ancestors in the quiet and uncontrolled possession of an ample domain, gradually sinking beneath our superior policy, our arts and our arms, have yielded their lands by successive treaties, each of which contains a solemn guarantee of the residue, until they retain no more of their formerly extensive territory than is deemed necessary to their comfortable subsistence. To preserve this remnant, the present application is made."[47] The denial of Cherokee claims does not rest on Marshall's refusal of empirical facts—on the victimization of a people "once numerous, powerful, and truly independent" but now culturally "sinking beneath our superior policy, our arts and our arms." Rather, it comes to rest on the legal question of whether the Cherokee Nation as a "domestic dependent nation" has standing to sue in the US Supreme Court as a sovereign state. Marshall asks, "Before we can look into the merits of this case, a preliminary inquiry presents itself. Has this court jurisdiction of the cause?"[48] According to the chief justice, they do not. The cries of the Cherokee cannot be heard; the violence they suffer cannot be recognized legally as violence and therefore cannot be redressed or repaired. Marshall's legal wizardry thus invents, as Barker concludes, "a sovereignty for indigenous peoples that was void of any of the associated rights to self-government, territorial integrity, and cultural autonomy that would have been affiliated with it in international law at the time."[49]

Constitutional law scholar Alexander M. Bickel points out in the context of *Dred Scott v. Sandford* (1851), a later Supreme Court ruling concerning whether a slave who travels from a slave to a free state might have legal standing to sue in court for his freedom, "It has always been easier, it will always be easier, to think of someone as a noncitizen than to decide he is a non-person, which is the point of the *Dred Scott* case."[50] Nonperson—that is, nonhuman—is precisely the point of *Cherokee Nation* as well. Here the problem of reparations and the human captures precisely this dynamic of injury and repair undeserving of political consideration. Seen from the perspective of the New World, reparation does not serve as a corrective either to the excesses of royal despotism or to the immoderations of liberal rights, dispossession, and accumulation under racial capitalism. It serves neither justice nor morality; rather, it positions the colonizer as a defensive Adversary and traumatized victim by managing a history of colonial conquest and death through a massive divestment of both Native lands and life. In the process, it forecloses the possibility of a critique of the *racial* state and *racial* reparations in the Transatlantic, a problem I will continue to examine in the following chapter on the Transpacific.

Locke's two bodies represent a story of not only political but also psychic disavowal and projection—indeed, a tale of social and legal rationalization embedded in an evolving European history of consciousness. The trope of the aggrieved Adversary—of the traumatized victim—is precisely where the psychic life of the human being, human understanding, and human consciousness enters Locke's writings on states of injury and repair. Nearly two and a half centuries later, the psychoanalyst Melanie Klein offers an account of reparation in a history of European *unconsciousness* that draws attention to how anxiety and guilt frame the compulsion to reparation by constructing a world of good and bad objects, human and inhuman subjects, across a similar colonial and Indigenous divide.

The Psychic: Klein's Colonial Object Relations

Together with the idealization of certain people goes the hatred against others, who are painted in the darkest colours. —MELANIE KLEIN, "Love, Guilt and Reparation"

If there is little scholarship in political theory focusing on the concept of reparation and its relationship to *colonial* conquest in Locke, there is even less critical attention paid to the colonial dimensions of reparation in Klein. As an aggrieved victim of Indigenous assault, the colonial settler not only monopolizes violence and repair under the sign of traumatized humanity but also indexes an emergent form of modern European subjectivity and consciousness. In his widely circulated *An Essay concerning Human Understanding*, which first appeared in 1689 almost simultaneously with his anonymous publication of the *Two Treatises*, Locke outlines his notion of modern subjectivity and consciousness based on empirical self-reflection.[51] Literary critic Cathy Caruth observes that "Locke's empiricism takes the form of a narrative, a narrative that tells a very specific tale about itself."[52] Locke's narrative offers us an early psychological portrait of the possessive individual as a defensive subject—as "a body worth defending," to borrow a term from the critical theorist Ed Cohen. As Cohen reminds us, the concept of "self-defense" emerged as a juridical concept in the mid-seventeenth century, during the English Civil War, when Thomas Hobbes defined it as the first "natural right."[53] Hobbes's observation provides one historical avenue from which to build a psychic account of reparation in relation to both self-consciousness and self-defense.

Before turning to my reading of reparation in Klein, I would like to return for a moment to *Beyond the Pleasure Principle* and the critical analysis I began in the introduction. In chapter 2, Freud departs from the "dark and dismal subject of the traumatic neurosis" of military combatants to examine the "*normal* activities" of children's play.[54] Seemingly far removed from the World War I battlefront, Freud nonetheless discovers "war in the nursery."[55] Observing the play of his eighteen-month-old grandson, Freud comes to witness how the death drive manifests in the infantile psyche, overshadowing the very beginnings of psychic life. He describes the young boy's development of a game—the famous *fort-da* episode—to cope with the comings and goings of his mother:

> This good little boy, however, had an occasional disturbing habit of taking any small objects he could get hold of and throwing them away from him into a corner, under the bed, and so on, so that hunting for his toys and picking them up was often quite a business. As he did this he gave vent to a loud, long-drawn-out "o-o-o-o," accompanied by an expression

of interest and satisfaction. His mother and the writer of the present account were agreed in thinking that this was not a mere interjection but represented the German word "*fort*" ["gone"]. I eventually realized that it was a game and that the only use he made of any of his toys was to play "gone" with them. One day I made an observation which confirmed my view. The child had a wooden reel with a piece of string tied round it. . . . What he did was to hold the reel by the string and very skillfully throw it over the edge of his curtained cot, so that it disappeared into it, at the same time uttering the expressive "o-o-o-o." He then pulled the reel out of the cot again by the string and hailed its reappearance with a joyful "*da*" ["there"]. This, then, was the complete game—disappearance and return. As a rule one only witnessed its first act, which was repeated untiringly as a game in itself, though there is no doubt that the greater pleasure was attached to the second act.[56]

Freud is at first perplexed by his grandson's behavior: the tossing away of the wooden reel, which the psychoanalyst interprets as representing the absent mother, is restaged with far greater frequency than its joyful return. Freud comes to parse this seemingly paradoxical act of repeated disappearance and *unpleasure* from several perspectives. First, it marks a *beyond* of the pleasure principle already evident in infantile life. Second, it underscores a psychic acquiescence to the reality principle, a renunciation of instinctual satisfaction that Freud characterizes as the child's "great cultural achievement."[57] Finally, it suggests a transformation of the subject in relation to the missing object insofar as his grandson seeks to transmute his unpleasant experiences of loss through the assertion of mastery. Characterizing this transformation as "defiant," Freud tells us that the young boy moves from a passive to an active relationship with the absent mother: "All right, then, go away! I don't need you. I'm sending you away myself."[58]

It is at this point in his discussion that "war in the nursery" takes on yet another permutation. Freud observes that a year later his grandson would often take the same toy and in anger throw it on the ground, exclaiming "Go to the front!" Freud explains, "He had heard at that time that his absent father was 'at the front,' and was far from regretting his absence; on the contrary he made it quite clear that he had no desire to be disturbed in his sole possession of his mother." Here Freud interprets the wooden reel's proliferating associations with an enlarged circle of objects through an Oedipalized framework that segregates love and hate, dividing them between two objects, between beloved mother and rival father. Many children express, he tells us, "similar hostile impulses by throwing away objects instead of persons."[59]

The exiling of the spindle to "the front" as a paternal foe explains why the defiant son would choose not to retrieve the wooden reel, consigning it to a dark and precarious fate in the killing fields beyond the domestic borders of the nursery. At the same time, this scene of exile to the front also presents us with a more general problem of *difference* in objects that exceeds traditional Oedipal interpretations. In the context of total war, which permeates the nursery through this imperative, the child's exclamations raise the urgent question of why some objects are thrown away *and* retrieved while others are not. "Go to the front!" forces us to rethink not just the psychic but indeed the social parameters of psychoanalysis in the face of an unrelenting global history of war, violence, and genocide binding together Europe, the Americas, Africa, and Asia under colonial modernity.

The evolution of the European liberal human subject and its storied history of consciousness might be said to meet its psychic limit in Freud's account of trauma and an unremitting death drive. This fracturing of liberal subjectivity and consciousness is a psychic dynamic that Klein explores through her classic theory of reparation as the modulation of psychic negativity and paranoid projection. If this Freudian scene of tossing away and retrieving anticipates the conceptual framework of Klein's account of object relations, whereby an aggressed and injured object is reclaimed in order to restore and repair it, then what is the logic of colonial object relations that leave certain objects unretrieved, left to perish in the dark regions beyond the circle of love and reparation?

It is no small irony that the Austrian-born Klein, who started her psychoanalytic training in Budapest amid the bombs of World War I, began to develop her theories of love and reparation in the interwar years, as the dream of enlightenment dissolved into the nightmare of total war and genocide.[60] From this historical angle, we might describe Klein's theory of reparation as an attempt to provide a new language for repair in order to rescue a besieged European liberal human subject in the midst of destruction. "The significance of the phantasies of reparation is perhaps the most essential aspect of Melanie Klein's work," the British psychoanalyst Joan Riviere observed in 1936. "For that reason her contribution to psycho-analysis *should not* be regarded as limited to the exploration of the aggressive impulses and phantasies."[61]

In the past decade, the turn to affect studies in the humanities has not only extended this discussion on the troubled relationship between violence and repair but also brought renewed attention to object relations in psychoanalysis as a theory of the emotions par excellence, supplementing the more language-based approach of Jacques Lacan and French poststructuralism. Klein's contemporary commentators, like Riviere before them, have often described her concept of reparation as largely a psychic roadmap for the preservation of love, albeit one born out of the hate and the psychic negativity of infantile life. In the same breath, her theory of reparation has also provoked resistance among theorists of childhood development precisely because of the ways it decidedly refuses any notion of the innocent child.

Indeed, Klein attributes to the pre-Oedipal infant an unrelenting aggression that makes her descriptions of infantile life difficult to stomach. In other words, Klein situates the emergence of Freud's death drive, often described as the speculative vanishing point of violence and negativity in psychoanalytic theory, not in adulthood but in the very beginnings of infantile psychic life.[62] In doing so, she effectively centers the death drive as the precondition for the emergence of the ego itself. If the *fort-da* game already renders the idea of the infant as a pure pleasure seeker dubious, once Klein centers the death drive as essential to processes of psychic being and meaning, "the pleasure/reality dichotomy is no longer pertinent."[63] It is eclipsed by the problem of aggressivity in psychoanalysis, one overshadowing the development of subjectivity from its very inception. Sedgwick observes that an "almost literal-minded animism makes Kleinian psychic life sound like a Warner Bros. cartoon . . . far too coarse-grained, [and] too unmediated."[64]

Yet Sedgwick, perhaps more than any other critic of her generation, has assiduously worked to gather and to recuperate Klein's coarse animism into a hermeneutics of what she describes as "reparative reading practices." Sedgwick argues that unlike paranoid reading practices—totalizing forms of inquiry that are anticipatory, self-reflexive, and coextensive with their criticism—reparative reading practices leave open an interpretative space for other affective and cognitive possibilities.[65] They offer alternative ways of negotiating negativity and surviving loss, especially in the historical wake of the AIDS pandemic to which Sedgwick is directly responding. As a mode of interpretation, a politics of survival, and a philosophical approach to the production of the liberal human subject, reparation represents for Sedgwick a new strategy for redressing histories of violence as well as an inventive mode for resisting social erasure. From this perspective, the preservation of love and life in Klein is often described as an ethical detour of Freud's death drive that allows intersubjective relations under social threat to persist and endure. The psychic process of "making reparation," Klein asserts, is a "fundamental element in love and in all human relationships."[66]

But what are the psychic limits of reparation? If we can describe reparation as a continuous psychic process modulating the negativity of intersubjective relations, then how might we understand the management of aggression as unevenly distributed and received among different objects and subjects in Klein? If the resolution of aggression in psychoanalytic theory is narrated as its conversion into morality and conscience through the superego in Freud, or its management through love in Klein, where exactly do aggression and hate go in these theories of affective sublimation? Reparation, as feminist psychoanalytic scholar Jacqueline Rose points out, has "often come to serve in the Kleinian corpus as a solution to difficulties—of negativity, causality, and knowledge."[67] As such, how might we evaluate the ethical complexities of this psychic transaction, this psychic leap of faith in liberal reason? How are love and hate differentially deployed, recognized, and socialized to circumscribe and to configure various object relations? Even more, how do love and hate *produce* good and bad objects in the first place—victims and perpetrators, humans and inhumans, valued and devalued objects worthy and unworthy of repair? How, we might ask, does affect become an object?

Reparation outlines the psychic process by which the infant negotiates a way out of the depressive position by making "good the injuries which [it] did in phantasy [to the mother], and for which [it] still unconsciously feel[s] very guilty."[68] According to Klein, because the mother is the original source of sustenance, her absence is registered by the frustrated infant through the splitting of the breast into good (available) and bad (unavailable) object. While the good object is loved and idealized, the bad object is hated and aggressed. The primal violence of splitting—what Klein calls the "paranoid position" of persecution and defense—destroys the mother as a hated object, but it simultaneously gives rise to an acute anxiety that the mother as a loved object has also been destroyed in the process. As the infant attempts to come to terms with the psychic havoc it has wreaked, the depressive position emerges as a response to the guilt arising from psychic destruction and anxiety—and from its nascent sense of responsibility toward the objects it has aggressed. If the infant can successfully negotiate this guilt, the reparative process ensues. Klein defines reparation as an act that encompasses "a variety of processes by which the ego feels it undoes harm done in phantasy, restores, preserves, and revives [dead] objects."[69]

By successfully mitigating psychic violence, the infant thus "repairs," "reinstates," and "restores" the mother as a separate object—indeed, a separate subject with agency and will. In this manner, the infant initiates an object relation not only with the mother but also with the rest of the world beyond her and the many other creatures in it. From this perspective, we might describe reparation as the psychic condition of possibility for the emergence of both the mother and the social world. Reparation is a notably one-sided affair: the infant repairs the mother not because of social coercion, forced submission, or even retaliation on the part of the aggressed mother. Rather, reparation is compelled by the infant's unconscious guilt over the phantasmic destruction of the object, by its acute anxiety concerning the loss of ties that bind it to the mother, and by "an internal mechanism that seeks to save objects from the ego's own destructive possibilities."[70] In short, subjectivity in Klein, as the historian Eli Zaretsky observes, "is inseparable from the idea that one has harmed or damaged the internal object on which one depends."[71]

Klein's account of reparation as the continuous psychic process of shuttling between different psychic positions (paranoid, depressive, reparative) is distinctive from Freud's more stagelike notions of childhood development (oral, anal, phallic, genital). Furthermore, it also diverges from Freud's influential account of melancholia, which narrates the incorporation of a lost object (the Oedipal father) as instituting a division between internal and external worlds with the emergence of a superego that inculcates the child into a past and into history through the generational transmission of morality and values.[72] In contrast, as the philosopher Judith Butler argues, Klein's "emphasis on preserving the object, which prefigures the work of reparation, suggests that incorporation also has its benevolent and ego-sustaining dimension, one that Freud overlooks in the effort to link melancholia with the institution of the super-ego."[73] Thus, despite her graphic insistence on the formative role of persecuting objects, Klein "also insists that the early internationalization of 'good objects' is both possible and necessary to insure later experiences of mourning that exceed their melancholic conclusions."[74] If Klein presents us with a theory of reparation in which instinctual anxiety and the pre-Oedipal death drive form "rudimentary knowledge" for the infant's subsequent experiences, it is equally important to emphasize that reparation marks the "child's phantasy not necessarily the object's need."[75]

To put it otherwise, the repairing of the object on which one depends through the preservation of love and the institution of intersubjective relations might be described as prefiguring any moral framework per se. The advent of guilt that impels the reparative process appears to emerge on the psychic scene, Butler observes,

> not in consequence of internalizing an external prohibition, but as a way of preserving the object of love from one's own potentially obliterating violence. Guilt serves the function of preserving the object of love and, hence, of preserving love itself. What might it mean to understand guilt, then, as a way in which love preserves the object it might otherwise de-

stroy? As a stopgap against a sadistic destruction, guilt signals less the psychic presence of an originally social and external norm than a countervailing desire to continue the object one wishes dead. It is in this sense that guilt emerges in the course of melancholia not only, as the Freudian view would have it, to keep the dead object alive, but to keep the living object from "death," where death means the death of love, including the occasions of separation and loss.[76]

This analysis of the relationship between guilt and moral prohibition raises a key problem concerning rudimentary knowledge outside history and memory: namely, the origins of morality and social constraint against violence that bind and channel the infant's negativity and aggression for the preservation of love and life in the reparative process. If "the question of survival *precedes* the question of morality," if Klein's infant desires the mother to survive only so that it may also survive, then "guilt—which is so often seen as a paradigmatically human emotion, generally understood to engage self-reflective powers and so to separate human from animal life—is driven less by rational reflection than by the fear of death and the will to live."[77]

While the infant's desire for self-preservation is threatened by its destructiveness toward the mother, this destructiveness is checked by the infant's utter dependency on her. In turn, the emergence of guilt and morality in this psychic scene might be described as the instrumental consequence of such a desire. Butler's provocative reading of the moral genealogy of reparation in Klein configures the mitigating of destruction and the preserving of love less as a concerted self-reflection and remorse over wrongdoing and aggression on the part of the infantile ego than as a consequence of the anxiety and guilt that the obliteration of the loved object would leave the ego "without the possibility of attachment and, hence, destroy[] the ego as well."[78] In short, the ego repairs the object on which it depends for its own survival, and in this way the infant's vulnerability takes precedence over that of the (m)other. Departing from Butler's insights, morality and self-reflection are therefore not the *causes* but, rather, the *effects* of reparation in Klein.

My reinterpretation of Klein proleptically dissociates reparation from morality and justice. If Klein delineates a psychic process by which the death drive does not hold ultimate sway—a theory of guilt and reinstatement of the aggressed object in which eros is not finally or fully extinguished by thanatos—the preservation of both love and life in Klein cannot finally be described as an ethical detour of the death drive. It is neither an antidote to aggression nor a synthesis of guilt into a more comprehensive and self-reflexive moral framework. It is neither an ethical assumption of responsibility and contrition toward the other nor the emergence of reparation under the sign of justice. In the end, although reparation for Klein constitutes "a fundamental element in love and in all human relationships," it is a decidedly fraught process of ethical complexity that a growing body of scholarship on Klein and the problem of "manic reparation" investigates.[79] As such, reparation might be rethought as an optative gesture, a leap of faith beyond liberal reason's moral ambit, a scramble to preserve and sustain life in the face of continued violence and negativity.[80] Ultimately, reparation must be approached as a kind of ethical responsibility that, as Butler concludes, is "bound up with an anxiety that remains open, that does not settle an ambivalence through disavowal, but rather gives rise to a certain ethical practice, itself experimental, that seeks to preserve life better than it destroys it. It is not a principle of non-violence, but a practice, fully fallible, of trying to attend to the precariousness of life, checking the transmutation of life into non-life."[81]

To put it another way, we might consider the genealogy of morals in reparation as the aporia of liberal reason. Reparation as a psychic process produces morality not as a rational self-reflection but, rather, as an irrational "fear of death and the will to live." If so, how might this aporia be explained, like in Locke, as a function of a colonial history of war and violence? How is aggression managed in this psychic scenario of morality as self-preservation and dependency? Who must bear the negativity of precarious life produced out of this scene? If morality and self-reflection are indeed the effects rather than the causes of reparation, then how might we think a history of colonial object relations in which morality and virtue are the results of this psychic transaction, in which a *colonial* morality and superego emerge as the consequence of these formative psychic resolutions?

To pose these questions from another angle, we might consider more carefully beyond the *fort-da* prototype when separation from a lost object produces anxiety, rage, guilt, mourning, and repair—and when it does not.[82] For reparation to occur, there must be an awareness of loss, a feeling of guilt, an apprehension of dependency; but what of the loss of an object that does not result in such affective acknowledgments? In this regard, reparation in Klein might be said to institute a dividing line between the good and the bad itself. Indeed, it retroactively constitutes a field of good and bad objects precisely through the management of love and the regulation of hate ordered by a long history of colonial modernity and liberal reason stretching back to Locke. If reparation stems from the instinct to survive and to preserve one's status as colonizer, it also becomes a psychic and social process of determining which object must remain "bad" and thus unsuitable for repair. From another perspective, the process of determining who is deserving of redress and what is a site of injury invariably raises the question of who has the authority and right to make such judgments. Ultimately, it raises the vexing theoretical dilemma in Klein of how to repair not a good but a *bad* object.

In her influential initial treatise "Love, Guilt and Reparation," Klein provides a psychic roadmap on the distribution of love *and* hate that creates a framework of colonial morality in its wake. The psychic management of affect in Klein can be described as the socialization of love and hate, which produces a realm of "proper" object relations for the self-possessed European liberal human subject—a chain of good objects from mother to others. The emotional extremity of infantile life, the polarity between love and hate, initially assumes its volatility because love and hate are projected onto one object: the mother. Klein narrates reparation as a psychic process of human development meant to dampen the affective charge of love and hate by segregating these extreme emotions through the continuing expansion and sorting out of the infant's social world. As the infant's social circle enlarges, love and hate are dislodged from the singular figure of the mother to be divided among *different* objects.

Within the extended space of family and kinship relations, siblings provide the first expansion of this dynamic, in which love for the mother displaces hate for her through its projection onto brothers and sisters in the form of sibling rivalry. (Notably, for Freud the displacement of hate in the *fort-da* episode is directed toward the Oedipal figure of the father, but Klein's pre-Oedipal psychic account moves us in an alternative direction.) In subsequent years, the classroom provides an even more ideal environment for this evolving segregation of love and hate. School life, Klein observes, "also gives opportunity for a greater separation of hate from love than was possible in the small family circle. At school, some children can be hated, or merely disliked, while others can be loved."[83] Unlike for brothers and sisters residing under one roof, Klein observes, it is socially permissible to love certain schoolmates and to hate others unconditionally.

This dynamic of affective segregation expands from the classroom to encompass a greater range of other subjects, including those from the older generation: "As the child grows to adolescence, his tendency to hero-worship often finds expression in his relation to some teachers, while others may be disliked, hated or scorned. This is another instance of the process of separating hatred from love, a process which affords relief, both because the 'good' person is spared and because there is satisfaction in hating someone who is thought to be worthy of it."[84] As good and bad objects become more clearly differentiated, reparation becomes less fraught insofar as "the subject's hate is directed rather against the latter [bad object], while his love and his attempts at reparation are more focused on the former [good object]."[85] The projection of hate onto "people not too close oneself also serves the purpose of keeping loved people more secure, both actually and in one's mind. They are not only remote from one physically and thus inaccessible, but the division between the loving and hating attitude fosters the feeling that one can keep love unspoilt."[86]

The developmental leap from school to the wider social world, Klein suggests, is marked by the reparative process assuming increased forms of abstraction—for instance, through the transformation of an idealized love for "mother" into an idealized notion of "motherland." The division and projection of love and hate as separate affects associates them not only with different persons but also, and more abstractly, with different places and things in the form of ideals. Thus, the abstraction of loved ones into an idealized set of interests and activities (such as nationalism) allows us to "speak of our own country as the 'motherland' because in the unconscious mind our country may come to stand for our mother, and then it can be loved with feelings which borrow their nature from the relation to her."[87] This abstraction and idealization tracks how a specific moral framework and a conventional set of collective social values and histories takes shape and form as an effect of evolving reparative processes.

Importantly, the transformation of love for the "mother" into love for the "motherland" allows us to apprehend liberal reason as the rationalizing of the colonial project. In other words, this transformation provides a psychic account of the simultaneous limitation and authorization of violence against a realm of objects retroactively constituted as good *liberal* and bad *Indigenous* objects. Indeed, we might go so far as to say that the object chosen for repair is not only constituted as good and worthy of reparation but also psychically constituted as *human*. In effect, those to whom repair can be offered become the very sign of the human—of value and valuable life in which intersubjective relations might be invested, cultivated, and sustained. Here it might be useful to consider how repairing what we value in relation to Locke's tenets of property, labor, and racial capitalism overdetermines not only the psychic constitution but also the psychic disavowal of a series of objects on which we depend but over which we assert independence, in the restricted social economy described by Klein.

If the splitting of good and bad objects, as the Polish British psychoanalyst Hanna Siegel argues, "orders the universe of the child's emotional and sensory impressions and is a precondition of later integration," we must consider how the faculty of social and psychic discrimination emerges here in relation to the continued differentiation between good and bad objects.[88] As such, I argue that reparation in object relations does not just outline a theory regarding the preservation and salvaging of love between subjects of violence; equally so, it delineates a process by which violence and hate is preserved *in potentia* for the continuous psychic and political consolidation of European liberal human subjectivity in a long history of colonial conquest and disavowal. Klein is perhaps ahead of herself when she notes, "Together with the idealization of certain people goes the hatred against others, who are painted in the darkest colours."[89]

While this reading of colonial objects relations in Klein supplements my analysis of reparations in Locke, I also would like to consider how Klein diverges from Locke's account of injury and repair as we move from the social to the psychic realm and back. Let me conclude, then, by turning to a final passage from "Love, Guilt and Reparation." Here we witness what happens when reparation extends beyond the charmed dyad of the mother-child and the larger family circle, into the space of the classroom, and finally into a wider social world defined by histories of colonial conquest. At the end of the essay, Klein unexpectedly interrupts her focus on an infant's attitude toward his mother to explore reparation from the perspective of a colonizer. In a section titled "Wider Aspects of Love," Klein, like Freud, brings war directly into the nursery, observing that the little boy's fantasies of "exploring the mother's body . . . contribute to the man's interest in exploring new countries."[90] Describing the European exploration of the New World as a simultaneous wish to run away from the mother while discovering her anew in an alternate location as "motherland," Klein offers this startling account of reparation:

> The child's early aggression stimulated the drive to restore and to make good, to put back into this mother the good things he had robbed her of in phantasy, and these wishes to make good merge into the later drive to explore, for by finding new land the explorer gives something to the world at large and to a number of people in particular. In his pursuit the explorer actually gives expression to both aggression and the drive to reparation. We know that in discovering a new country aggression is made use of in the struggle with the elements, and in overcoming difficulties of all kinds. But sometimes aggression is shown more openly; especially was this so in former times when ruthless cruelty against native populations was displayed by people who not only explored, but conquered and colonized. Some of the early phantasied attacks against the imaginary babies in their mother's body, and actual hatred against new-born brothers and sisters, were here expressed in reality by the attitude toward the natives.

The wished-for restoration, however, found full expression in repopulating the country with people of their own nationality.[91]

This is an astonishing passage. In moving beyond the mother-child dyad into the psychic dynamic of sibling rivalry, and finally into the social history of European colonialism, reparation as the possibility of an experimental ethics is put into explicit crisis. Reparation emerges here as neither anxiety nor ambivalence toward an injured other. To the contrary, it functions as a disavowal of responsibility in a history of colonial war and violence that preserves and extends life to some while simultaneously withholding it from others.

In the face of radically asymmetrical power relations, reparation breaks down. Klein's infant aggresses against the mother on whom it is utterly dependent, eventually coming to check its sadism through the production of guilty feelings and the preservation of love whose moral status remains in ethical abeyance, if not ethical possibility. Here Klein aligns the European colonizer with the helpless infant, although he is, in actuality, less the powerless infant "phantasi[zing] attacks against the imaginary babies" than the powerful mother, wreaking real violence against Native populations. The colonizer's material destruction of the Indigenous in "former times" does not bring about a state of heightened anxiety and guilt, a mitigating of violence, or a restoration of the Native other. Instead, reparation comes to name the psychic process of responding to European colonization and genocide of Indigenous peoples by repopulating the New World with images of the self-same.

Incredibly, a long history of Indigenous dispossession and death is psychically configured as "restoration." As Klein asserts, the "wished-for restoration" is given not to the Native other, the actual victim of colonial violence "in reality," but to the European colonizer in the form of "repopulating the country with people of their own nationality." Thus, we witness in this passage the splitting of Klein's theory of reparation, like that of Locke: inside the motherland, reparation preserves life and love for the self-same as the production of a good liberal object worthy of repair; outside the motherland, it authorizes violence toward the Native other as the production of a bad colonial object unworthy of consideration. This moment marks an aporia of liberal reason in the reparative process—a leap of psychic faith. Violence is quelled and aggression is resolved only through a constitutive substitution of the self-same subject and disappearance of the figure of the violated Native. This disappearing act is marked by the simultaneous emergence of a colonial morality as a force not constraining genocide but, rather, sanctioning it.

We might say, along with the performance studies scholar Joshua Chambers-Letson, that Klein glosses over in this particular passage one of the primary components of her own theory: namely, the guilt born of "ruthless cruelty against native populations."[92] Instead, she replaces the proper subjects of European aggression—the vanquished Natives—with "new-born brothers and sisters" of the European tribes. While Klein's condemnation of colonial aggression is conveyed simply through the phrase "ruthless cruelty against native populations," whatever moral status this condemnation might carry is immediately rendered unstable immediately by the following sentence: "The wished-for restoration, however, found full expression in repopulating the country with people of their own nationality."

The instability of this "however" is worth contemplating, as it marks both a convergence and divergence from Locke's political account. On the one hand, the "however" might be interpreted as symptomizing the historical reality of European colonialism in which reparation as the redistribution of life and property was in fact given to the colonizer rather than the Native. On the other hand, this "however" might also be seen as (unconsciously) indicating the reparation that *should* have been offered to the Natives—a glimmer of ethical contrition rather than an instance of psychical defense. In either case, Klein does not grapple with this willing away of ambivalence—that is, with the ethical implications of this segregation of affect, the sorting out of love and hate between the colonist and Native, or the production of a colonial morality that allows the colonial perpetrator of violence to assuage his anxiety and guilt.[93] Indeed, we might describe this moment in Klein as a detailed account of the psychic production of bad-faith liberal white guilt.

The irresolution of the "however" as psychic symptom thus marks the injustice of the historical situation, but it also functions as a heuristic for the deconstruction of liberal reason. It provides the insight that the emergence of the morality in Klein is specifically the production of a *colonial* morality, a *colonial* superego. This "however" also lends great historical irony to Klein's statement that "the process of displacing love is of the greatest importance for the development of the personality and of human relationships; indeed, one may say, for the development of culture and civilization as a whole."[94] Echoing the renunciation of instinctual satisfaction on the part of Freud's grandson, this psychic sleight of hand is the "great cultural achievement" of a consequential social renunciation, of what I am describing in this chapter as the political and psychic genealogy—indeed, the political and psychic unconscious—of colonial object relations. From this perspective, we might say that the displacement of love does not resolve but, to the contrary, engenders an intersubjective crisis between colonizer and colonized.

In Klein's account of New World dispossession, reparation is ultimately driven by the colonizer's will to life and self-preservation in a new and unyielding territory under the banner of a colonial morality thoroughly divested of historical response and responsibility toward the Native other. His precarious life decidedly trumps the precarious existence of the Native other. Reparation cannot be described here as an ethical detour of the death drive. More accurately, reparation comes to circumscribe love, conditioning both nation and nationalism. It constitutes the colonial settler as the idealized lover of and for the European family of nations. Reparation thus names the collective social and psychic processes by which love becomes a naturalized property of the European liberal human subject, foreclosing in the process any possibility for racial reparation for the violated Native.

Similar to Locke, the figure of the Indian appears in these closing pages of "Love, Guilt and Reparation" only to disappear, written out of Klein's psychic account of reparation. In its place, the European colonizer monopolizes *both* sides of the psychic equation: he is both *perpetrator* of violence and traumatized *victim* deserving of repair, short-circuiting legal notions of trauma and injury, of reparation and human rights, which demand a clear distinction between these terms in law and politics, a point I will elaborate on in the next chapter. In the process, we witness the psychic consolidation of proper boundaries not only of the liberal human subject but also of European family, kinship, and nation as a closed circuit of injury and repair. Klein supplements Locke's account of reparation by not configuring it as the restoration of invaded territory or redress for colonial violence against the Native other. Rather, reparation in Klein delineates the redistribution of love, life, and land—the arrogation and naming of conquered Native territory—as "motherland."

As such, Klein's colonial object relations cannot be thought of as a moral response to or concerted self-reflection on violence. It cannot prefigure any synthesis of guilt into the ethical assumption of historical response and responsibility—namely, the impulse toward justice. Instead, it must be described as kind of "moral sadism," the rationalizing of the colonial project across political and psychic registers of liberal reason, or "as a mode of persecution that passes itself off as a virtue."[95] Drawing on this notion of persecution as a root of bad-faith liberal white guilt allows us to theorize how reparation in Klein might be understood as "a racialized performance of passivity," to borrow a concept from the African American theorist Amber Musser, or as a liberal "care of the self" dissociated from its violent colonial past, to reconfigure a concept from Michel Foucault.[96] As a form of negative narcissism, this bad-faith gesture works against responsiveness as it "recoils from the other, from impressionability, susceptibility, and vulnerability," in the words of Butler.[97] In short, liberal white guilt eschews ethical responsibility toward the injured Native precisely by psychically colonizing its suffering.

Rose observes that if "psychoanalysis has something to say about war, one might also reverse the proposition and suggest that war has something to say about psychoanalysis, or at the very least about its own relationship to knowledge, its own conception of what constitutes the truth. War does not only appear as an object of psychoanalytic investigation, of course. It provides the living context for key moments in the history of psychoanalysis."[98] In the wake of total war—and, even more importantly for our discussion, a long history of colonial dispossession and death in the New World—injury and repair cannot be described as oppositional, nor can aggression and morality. Rather, these terms must be considered as constituting a larger psychic and political dialectic underpinning the evolution of liberal subject outside universal norms of the human. As Zaretsky contends, reparation in Klein is "dominated by responsibility to particular others to whom one has incurred obligations, not in virtue of being generically human, as in Kant, but because one has found oneself in specific relations and circumstances."[99] While Rose speaks of war in general and Zaretsky focuses on specific ties that bind a subject to his chosen object, the story of colonial violence that I narrate here expands their insights to consider a larger, collective history of reparations and the human in New World conquest.

With Hegelian echoes of lord and bondsman, reparation in Klein is transformed into an alibi for war and aggression by displacing the actual, external violence of colonial settlement and genocide into an *internal* struggle of European family and nation. Similar to Locke's political account of the "innocent wives and children" of the European tribes, the "wished-for restoration" in Klein's psychic account of reparation is internalized as self-consciousness toward brothers and sisters. This sublation of aggression and alterity finds its "full expression," its true form, in a tale about the universal European liberal human subject and its storied history of consciousness. Reparation appears as the differential production of the human through the affective distribution of precarious life, as it constitutes and separates good objects deserving of care and redress from bad objects meriting no consideration. It divides the man of reason from the unreasonable brute, marking the limits of liberal reason as the very sign of our humanity.

If Locke and Klein might be said to be describing quite accurately the historical reality of British settler colonialism and Indigenous genocide in the

Americas, their theories of reparation analyzed here also mark a perverse type of colonial knowledge: the ghostly political and psychic processes of doing and undoing the human in the context of New World conquest. The political and psychic aporia of reparation in both Locke and Klein symptomize this problem of knowledge, a forgetting of a history of colonial violence and repair—its unrecognized, dead, displaced, and erased subjects. As such, any analysis of reparations and the human must occasion the investigation of the human not as a ground of, but as a *problem for*, ethics.

Like Locke's account of dispossession and genocide in America, unchecked and unrecognized aggression against Native populations in Klein is occluded in psychic life through a colonial morality that emerges as the psychic foundation for its continued renewal and persistence. Klein's theory of reparation, similar to Locke's, is grounded in colonial conquest, a remarkable fact largely ignored by her commentators. Locke's dissociation of colonizer from Aggressor-perpetrator and Native from Adversary-victim in the *Second Treatise* authorizes the massive redistribution of human life and property under colonialism. In turn, Klein's theory of reparation works to carve out a psychic terrain that manages and rationalizes the psychic negativity of this troubled colonial history in the name of repair even as it exposes, in the same breath, the construction of a colonial superego as one of its persisting effects. We are left with the problem of trying to narrate an alternative history of loss beyond repair, an alternative history of racial reparations and the human.

The devastation of genocide in Europe and nuclear holocaust in Asia instigated a new model for reparations and human rights in the postwar period. In the wake of unfathomable catastrophe and with the rise of the American Century, the reinvention of reparations and human rights under the shadows of Cold War conflict constituted, we are conventionally told, a break from the history of human precarity and the ongoing violence of the sovereign state. In contrast to these accounts, I suggest that war and violence in the Cold War Transpacific continues, rather than departs from, the social and psychic logic outlined here, and in the following chapter I turn to this region.

2

Beyond Trauma

War and Violence in the Transpacific

But with Hiroshima, where the continuity of life was, for the first time, put into question, and by man, the existence of any survivors is an irrelevancy, and the interview with the survivors is an insipid falsification of the truth of atomic warfare. To have done the atom bomb justice, Mr. Hersey would have had to interview the dead. —MARY MCCARTHY, "The Hiroshima *New Yorker*"

Let all the souls here rest in peace. For we shall not repeat the evil.
—MEMORIAL CENOTAPH, Hiroshima Peace Memorial Park

The Holocaust and the atomic bombings of Hiroshima and Nagasaki marked a radical shift in our conception of the human being and, more specifically, of human precarity whereby, in Mary McCarthy's sober assessment of the latter events, "the continuity of life was, for the first time, put into question, and by man."[1] In response, a new international order of human rights with attendant notions of reparations arose from the ruins of World War II. This new legal regime sought to subrogate the sovereignty of the nation-state in the hopes of defending the sovereignty of the individual. Traditionally, reparations were claimed by a victorious state over a vanquished one as compensation for the costs of war. For the first time in history, reparations were extended to encompass individual and group claims against state-sponsored violence in the name of human rights.

Together, reparations and human rights aspired to protect the "abstract nakedness of being human," to return to Hannah Arendt, beyond the striking failures of the modern nation-state to ensure the sanctity, indeed the very continuity, of human existence.[2]

Together, genocide and nuclear holocaust summoned the specter of planetary annihilation in the world imagination—a prospect made real by the advent of Cold War hostilities between East and West under the threat of mutual assured destruction. However, as I note in the introduction, the "Final Solution" and the atomic bombings also cleave from one another in significant ways. In the space of postwar Europe, the history of the Holocaust is settled: Nazis were perpetrators and Jews were victims. In contrast, in the space of postwar Asia, there was and continues to be little historical consensus as to who were the victims and who were the perpetrators. Although the United States deployed two atomic weapons on Japan, with an estimated range of 110,000–210,000 deaths, the status of those responsible for the bombings as well as those who perished from it remains indeterminate.[3] To this day, nuclear arms are considered legal instruments of warfare. Thus, unlike the Holocaust, where West Germany paid significant reparations to Jews as well as to the state of Israel, and where the Holocaust is excoriated as the epitome of "evil," cementing Germany's legal and moral status, the possibility of reparations for the atomic bombings remains unthinkable.[4]

Consider one illustrative example of this important but largely unremarked obfuscation of accountability for state-sponsored mass murder: the cryptic inscription on the Memorial Cenotaph in the Hiroshima Peace Memorial Park. It reads, "Let all the souls here rest in peace. For we shall not repeat the evil." Who is the "we," and what exactly is the "evil" that shall not be repeated? Does the first-person plural pronoun refer to Japan, to the United States, or to humanity in general? And does the "evil" encompass the singular event of the obliteration of Hiroshima by a US atomic weapon, a wider history of Japanese colonization and aggression in its so-called Greater East Asia Co-prosperity Zone, a longer legacy of Western imperialism in the region, or perhaps war and violence in general?[5]

The deliberately vague and open-ended possibilities for these referents reflect this unsettled history—the impossibility of determining who exactly are the victims and who exactly are the perpetrators in the aftermath of Hiroshima. In other words, lacking either a proper subject or a proper object of the evil, the epitaph raises the problem of agency and responsibility in the wake of atomic disaster. If the Memorial Cenotaph seeks to commemorate the many souls who perished in Hiroshima, accountability for these losses dims in the face of such regnant indeterminacies: we have numerous victims, but there are no clear perpetrators. How did this come to be?

Building on my analysis of reparations in political psychoanalytic theory under New World colonization, I turn here to the unexamined links between psychic and political approaches to trauma in the space of Cold War Asia and the postwar context of decolonization. Here I investigate the division between victims deserving and undeserving of repair both in the face of identifiable perpetrators and in their absence, from the Holocaust to Hiroshima, and from Nuremberg to the Tokyo War Tribunals. I argue that the arrogation of trauma and repair, along with the uneven nomination of victims and perpetrators, situated in different geopolitical spaces and times, carves out a privileged zone of exhausted and victimized humanity, with significant implications for addressing the injuries of violated human beings beyond Europe and the New World analyzed in chapter 1. The serial military actions, conflicts, and partitions that engulfed Cold War Asia in the aftermath of atomic destruction—in Korea and Vietnam, most notably—created an unending stream of displaced and stateless refugees through a series of proxy wars marking a global battle between capitalist and socialist modernity.

The first section of chapter 2, "The History of the Subject," begins with an examination of the relationship between psychoanalytic accounts of trauma and legal designations of victims and perpetrators as they emerge in the context of European genocide, the International Military Tribunals at Nuremberg (IMT), and what Joan Wallach Scott describes as "the judgment of history" arising from those formative events. Here I turn to Sigmund Freud's reading of Tancred and Clorinda in *Beyond the Pleasure Principle* by investigating intellectual historian Ruth Ley's critique of "unclaimed experience," a concept developed by Cathy Caruth, which helped to establish the interdisciplinary field of trauma studies. The second section, "The Subject of History," proceeds with a discussion of John Hersey's August 1946 *New Yorker* essay, "Hiroshima," detailing the aftermath of the atomic bombing through the eyes of six surviving inhabitants of the devastated city. In the wake of that catastrophe, as well as the corresponding International Military Tribunals of the Far East (IMTFE) modeled on the legal precedents established at Nuremberg, I consider how psychoanalytic approaches to the history of the traumatized subject supplement the subject of Cold War history still in search of an impossible historical consensus.

I conclude in the third section, "Human-Civil-Inhuman," by turning to the realm of literature. Unlike history and law, which demand the nomination of a singular victim and a singular perpetrator, literature can deconstruct this divide. At the same time, like psychoanalysis, literature can probe the social and psychic effects of the definitive separation of victims from perpetrators, and it can investigate the ethics of recognition in the face of legal and moral judgment. I analyze a number of novels by Kazuo Ishiguro (b. 1954) and Chang-rae Lee (b. 1965). Although the two authors are rarely considered together—they are associated with different Anglophone and Asian American literary traditions, respectively—both are intimately concerned with violence, witnessing, responsibility, and repair as they construct subjects of bad faith in the context of war-torn Asia.[6]

Together Ishiguro and Lee illustrate the centrality of the region to global rearrangements of sovereign power across the twentieth century. Ishiguro's work positions Asia in relation to a declining Europe waning in colonial dominance, while Lee's oeuvre situates Asia in relation to an ascendant United States during the American Century. In the process, their novels unsettle received notions of the human and the inhuman, and of the civil and civil rights, as they collectively shape traumatized subjects worthy or unworthy of repair. Ultimately, the two authors' fictional writings help us connect and rethink the postwar ascension of reparations and human rights in the aftermath of atomic destruction, Japanese American internment, and the enslavement of Korean comfort women not as a moral response to, but instead as a form of, continued state violence. Together Ishiguro and Lee unearth the possibilities and limits of what I describe as the problem of racial reparations and the human in the Transpacific.

The History of the Subject

I begin my investigation of the history of the subject by reminding us that psychoanalysis itself has a history, one embedded in colonial modernity. In recent years, various scholars in postcolonial studies and new European history have started to examine psychoanalysis from this perspective: in relation to the "anthropological" Freud; as a developmental discourse of European civilization versus its primitive others; in regard to the circulation of psychoanalytic theories, methods, practices, and practitioners between metropole and colony; and in terms of the rise of authoritarianism and fascism. Concomitantly, on this side of the Atlantic, a growing number of race scholars have considered how psychoanalysis provides a vocabulary for analyzing not only sexual but also racial desires and prohibitions constituted by legacies of colonialism, dispossession, slavery, and exclusion haunting the foundations of US law and society. In this section I advance a postcolonial critique of psychoanalysis by focusing on the history of the *traumatized* subject—how psychoanalytic and legal accounts of trauma converge and diverge to affirm particular victims and perpetrators while forgetting others who perished in the face of colonial war and violence.

Let us return to my initial reading of Freud from chapter 1 of this book. As we have seen, Freud first outlined his theory of the death drive in *Beyond the Pleasure Principle* (1920) in relation to patterns of unremitting trauma exhibited by military combatants returning from the World War I battlefront. "There exists in the mind a strong *tendency* toward the pleasure principle," he observes, "but that tendency is opposed by certain forces or circumstances, so that the final outcome cannot always be in harmony with the tendency toward pleasure."[7] Similar to hysterics, the soldiers Freud treated suffered from "reminiscences." Unlike these civilian patients, however, their reminiscences could be traced not to a primordial or pleasurable object of desire but, unexpectedly, to their recent participation in grisly violence on the battlefield. Witnessing the involuntary reliving of their painful experiences—what he came to diagnose clinically as "shell shock" and what later, following the Vietnam War, would evolve into contemporary paradigms of posttraumatic stress disorder (PTSD)—Freud wrote *Beyond the Pleasure Principle* to better understand the human drive beyond pleasure toward death.

Freud offers a series of contemporary examples drawn from daily life to illustrate this inexplicable compulsion toward a scene of pain—of the traumatized subject's reliving an agonizing incident rather than "*remembering* it as something belonging to the past."[8] These cases range from veterans returning from the World War I battlefront to survivors of industrial and railway accidents and even, as I discussed earlier, to his infant grandson's quotidian distress at the unanticipated departures of his beloved mother from the domestic space of the nursery—the famous *fort-da* episode. In chapter 3 of his book, Freud unexpectedly withdraws once again from the contemporary war-torn landscape around him to develop yet another aspect of the death drive drawn from the battlefields of historical fiction. Here he analyzes Torquato Tasso's *La Gerusalemme liberata* (1581), an early modern epic poem recounting the story of Tancred and Clorinda, two ill-fated lovers on opposing sides of the First Crusade.

Tasso's poem recounts Tancred's accidental slaying of his lover Clorinda during the liberation of Jerusalem (1099). For Freud, this tragedy flies in the face of the pleasure principle not only because of the killing itself but also because of the soldier's inadvertent repetition of its violence. He explains, "Its hero, Tancred, unwittingly kills his beloved Clorinda in a duel while she is disguised in the armour of an enemy knight. After her burial he makes his way into a strange magic forest which strikes the Crusaders' army with terror. He slashes with his sword at a tall tree; but blood streams from the cut and the voice of Clorinda, whose soul is imprisoned in the tree, is heard complaining that he has wounded his beloved once again."[9] Clorinda's demise is exemplary—especially painful and moving for Freud—insofar as the repetition of its violent ends seems initiated not by Tancred's active cultivation of the death drive but by his passive subjection to it. That is, it seems impelled by some sort of unfathomable fate to which the crusader is subjected beyond any conscious will or control: "We are much more impressed by cases where the subject appears to have a *passive* experience," Freud observes, "over which he has no influence, but in which he meets with a repetition of the same fatality."[10]

In *Unclaimed Experience*, Caruth famously analyzes this passage in Freud's book as precipitating a crisis of historical reckoning that troubles our understanding of the traumatized subject. Caruth's analysis of "unclaimed experience"—experience not consciously motivated yet subjectively felt—addresses both the historical and the ethical implications of a repetition compulsion that, as she observes, "exceeds, perhaps, the limits of Freud's conceptual or conscious theory of trauma."[11] While Freud highlights Tancred's unwitting repetition of violence against his beloved, Caruth emphasizes instead the enigmatic and sorrowful voice of the departed lover who cries out from the wounded tree. She thereby focuses our attention on the involuted relationship between psychoanalysis and history, on the problem of a traumatic history that arises precisely "where *immediate understanding* may not."[12]

The mystery of Clorinda's unanticipated cry marks a history of trauma that, according to Caruth, is "referential precisely to the extent that it is not fully perceived as it occurs; or to put it somewhat differently, that history can be grasped only in the very inaccessibility of its occurrence."[13] Tancred does not just inadvertently repeat his act of violence, she asserts; rather, "in repeating it, he for the first time hears a voice that cries out to him to see what he has done."[14] If Clorinda's recurrent demise and protests situate Tancred's death drive in a repetition compulsion decidedly beyond the pleasure principle and in excess of any semblance of conscious control, it also insists on an understanding of this tragedy as an unresolved quandary between the history of the traumatized subject and the subject of traumatic history. Tancred's predicament exceeds its hero in both comprehension and scope.

For Caruth, the story of Tancred is also the story of Clorinda: trauma marks "the enigma of the otherness of a human voice that cries out from the wound, a voice that witnesses a truth that Tancred himself cannot fully know."[15] Even more, as medieval scholar Matthew Aiello points out, it is at this precise moment of their heated exchange in Tasso's poem that the voice of Clorinda expands to encompass Christian and pagan alike, moving from the individual woman warrior to a much larger collective: "I was Clorinda," she tells the hero, "now imprisoned here / Yet not alone within this plant I dwell / For every Pagan lord and Christian peer / . . . But here they are confined by magic's spell."[16] In the final analysis, history is never simply one's own but, rather, as Caruth concludes, "precisely the way we are implicated in each other's traumas."[17]

To put it otherwise, Clorinda's cry links a history of war and violence that belongs as much to the surviving Tancred as to the departed Clorinda and, equally important, to the numerous others who perished in the religious battle and whose collective traces remain condensed in a haunting and imprisoned voice emanating from the dark forest. The agent of her destruction, Tancred thus simultaneously bears witness to Clorinda's survival—more precisely, to her living on in history as an enigmatic voice that remains stubbornly opaque to him. From this perspective, we might describe Clorinda's cry as a kind of ethical call arising from dead and injured subjects outside circuits of recognition and repair but who nonetheless demand response. Her cry demands the anguished hero to see and, even more, to take responsibility for what he has done. Moreover, it demands that we, in turn, devise a more complex accounting of the slaughtered enemy—indeed, those deprived of any remembrance at all. It requires, that is, a more complex understanding of historical response and responsibility engendered by the nomination of victims and perpetrators in the aftermath of violence, one produced across both psychic and political registers. To put it simply, who is the victim in this scenario between the ill-fated lovers, and who is the perpetrator?

Caruth's influential approach to the inverted relations between trauma and history has been met with both approbation and resistance. Leys, for instance, has been an especially vocal critic of Caruth. She contends that by positing a nonreferential theory of history, a history of trauma that escapes understanding at the moment of its occurrence, Caruth traffics in both historical and political relativism. In particular, Leys takes issue with Caruth's implication of Tancred as the privileged and "traumatized victim" of this violent encounter. She argues that if Tancred is a traumatized survivor of this battle, he should not in turn be allowed to claim the status of victim insofar as it is Clorinda who is the unfortunate object of his repeated and fatal aggressions. According to Leys, the murdered Clorinda is decidedly the victim, and the surviving Tancred is necessarily the perpetrator.

Leys interprets Caruth's case of mistaken identity as the effect of a wayward poststructuralist who, on the one hand, is committed to installing the Holocaust as the paradigmatic trauma of modernity but, on the other hand, wrongly associates the event with precipitating "an epistemological-ontological crisis of witnessing, a crisis manifested at the level of language itself."[18] For Leys, the political implications of Caruth's deconstructive project are dangerous if not execrable. By prioritizing the indeterminacy of language and communicability in her theory of trauma over the realpolitik of European genocide, Leys argues, Caruth excuses Tancred's gendered violence against Clorinda and, by extension, muddles victim and perpetrator in the historical context of Nazi and Jew. If "the murderer Tancred can become the victim of the trauma and the voice of Clorinda testimony to *his* wound," Leys concludes, "then Caruth's logic would turn other perpetrators into victims too—for example, it would turn the executioners of the Jews into victims and the 'cries' of the Jews into testimony to the trauma suffered by the Nazis."[19]

There is much to unpack in Ley's critique of unclaimed experience, and her will to moral certitude, judgment, and blame. Let me begin first by returning to Tasso's poem—to the fact that Clorinda is a Muslim, not a Jew, and that the violent religious conflict narrated by the poet Tasso concerns a holy war between a Christian crusader (Tancred) and a Muslim infidel (Clorinda).[20] Christians are insistently configured in Western history as victims of unwarranted Islamic aggression. This tradition of a defensive Western civilization emerges from the Crusades and binds the medieval to the early modern periods, from the Golden Horde to the Ottoman Empire to the Reconquista. In the Enlightenment, it transforms into what Carl Schmitt has described as a "political theology" of secularized Christian ethics. In our contemporary context, it appears as an interminable "war on terror" fixing a secular (Christian) West in opposition to a fundamentalist (Muslim) East—an abiding feature of Orientalism explored by the postcolonial critic Edward Said in his eponymous treatise. Seen from this long-standing European Christian perspective, describing Tancred as perpetrator rather than victim, or as enemy instead of friend, may not be as historically clear-cut as Leys would ideally desire.

In other words, to interpret Tancred's gendered aggression toward Clorinda solely as prefiguring the culminating violence of the "Final Solution," of the destruction of European Enlightenment under total war, would require an inquiry into how an extended history of Orientalism and religious-racial conflict becomes subsumed into an exclusive narrative about the dissolution of Western civilization and its history of consciousness through the Nazi extermination of Jews. Tancred's gendered violence can also be examined from an additional angle. To the extent that Tancred is traumatized, his psychic distress may be the result not of killing per se but, rather, the killing of a *wrong* object. In other words, Tancred's trauma may be the result of a case of mistaken identity. The

"Christian peer" believes that he has killed a "Pagan lord" when, in fact, he has killed his beloved Clorinda disguised as such.

Tasso's recounting of their tragic love story in *La Gerusalemme liberata* underscores the unthinkable and repeated violation of this human intimacy, of kith and kin. Clorinda's anguished voice draws immediate attention to the problem of a divided collective, to our uneven responses to the cries of *different* objects—indeed, to the cries of dead and injured subjects that often go unheeded. In the final analysis, the enigma of Clorinda's voice raises the urgent *ethical* problem of whose cries deserve recognition and whose cries remain unheard.

As they blur the divide between victims and perpetrators, Clorinda's exhortations simultaneously trouble the boundaries between friend and enemy, intimate and stranger, defining the dichotomized politics of war. Equally urgently, they force us to consider how the socialization of Tancred's psychic predicaments and the politicization of his death drive facilitate the legal arrogation of trauma and victimhood—the legal nomination of traumatized *human beings* deserving of repair and to what political ends. To put it simply, we do not necessarily repair injured human beings. Rather, it is those to whom we offer repair who *become* human by virtue of our recognition of them as such—personified by those who merit this concern as well as by those who render it. Here we encounter a refiguration of the closed circuit of recognition and repair, returning us to my analysis of Melanie Klein in chapter 1. Caruth's reading of Freud and the crisis of witnessing opens up a nonreferential theory of traumatic history for renewed investigation across both the political and psychic domains.

Without endorsing Ley's hyperbolic conclusions regarding Caruth's misrecognition of Nazi for Jew, and victim for perpetrator, what I would like to highlight in this exchange between the historian and the literary critic is precisely the slippage between legal and psychoanalytic accounts of trauma, and of the boundaries between victims and perpetrators established precisely through their cleaving. On the one hand, trauma in psychoanalysis is catholic: in the face of unremitting war and violence, anyone can potentially be traumatized. Unlike law, psychoanalytic theory makes no distinction between victims and perpetrators. Both victims and perpetrators alike can suffer from trauma's debilitating effects. As the Holocaust historian Michael Rothberg observes, the "categories of victim and perpetrator derive from either a legal or moral discourse, but the concept of trauma emerges from a diagnostic realm that lies beyond guilt and innocence or good and evil."[21] It behooves us to recall that Freud's paradigmatic victim of trauma in *Beyond the Pleasure Principle* is, in fact, the shell-shocked *German* soldier.

On the other hand, trauma in law is highly restricted: it is the property of victims. While trauma in psychoanalysis has a universalizing impulse and therapeutic intent, trauma in law is concerned precisely with its limited distribution. Indeed, we might say that through trauma's uneven allocations, law attempts to fix accountability and responsibility—cause and effect, guilt and innocence, blame and blamelessness. Ultimately, it seeks to contain the violence it both engenders and monopolizes as its sovereign right through the production of moral culpability. As such, the arrogation of trauma in law functions to nominate victims and, in turn, to designate perpetrators. In short, it creates a sustained boundary between these opposing subject positions precisely through trauma's pointed distribution in a politics of liberal rights and recognition. I will return to this fraught issue momentarily.

From this perspective, Ley's desire to fix Jew as "victim" and Nazi and "perpetrator" in the aftermath of genocide represents an effort to ensure what Scott has described as the "judgment of history." There is a "popular belief," Scott observes in the context of the Holocaust, "that there is a certain moral impeccability about history's judgment. It is a secular version of the biblical day of reckoning at the End of Times, serving the same phantasmic function, providing transcendent reassurance for one's moral positions."[22] In the face of unspeakable violence, the moral force of history—a moral force for which law purports to speak and by which law seeks to punish the evil deeds of mortals through a detailed historical accounting of their unspeakable crimes—is something we "cannot not want," to borrow a concept from postcolonial scholar Gayatri Spivak in her critique of human rights.[23]

Yet we should not confuse moral judgment with the problem of recognition. As Judith Butler writes, "I think that it is important, in rethinking the cultural terms of ethics, to remember that not all ethical relations are reducible to acts of judgment and that the very capacity to judge presupposes a prior relation between those who judge and those who are judged. The capacity to make and justify moral judgments does not exhaust the sphere of ethics and is not coextensive with ethical obligation or ethical relationality."[24] We might bring a similar critical insight to those who have the capacity to repair and those who are given repair. That is, we cannot conflate histories of legal judgment and reparations with an ethics of recognition, ethical relationality, or human reciprocity—an insight we also glean from our discussion of the infant's one-sided drive for reparations in Klein.

As Scott points out alluding to Karl Marx, history not only is made by humans, albeit under circumstances often beyond our own making, but also resists any final closure. Conceiving of history as an autonomous moral force bending toward the arc of justice underwrites an impossible narrative of universal human progress and reason. At the same time, it reinscribes the modern nation-state—the *victorious* nation-states that seeks to adjudicate crimes against humanity and mete out reparations and redress—as the privileged agent of that redemptive history. The conflation of the modern state with the foundations of law and history, Butler asserts, is "the temporal framework that uncritically supports state power, its legitimating effect, and its coercive instrumentalities."[25] And that modern state, as numerous postcolonial and critical race theorists have emphasized, and as the catastrophe of the Holocaust starkly underscores, is a racial state.

Scott observes that in our contemporary context, the Holocaust serves as the urtext for the judgment of history, stubbornly resistant either to extending its moral force or, as some critics engaged in postcolonial and psychoanalytic theory have argued, to sharing its monopolization of trauma with other catastrophic events outside non-Western contexts.[26] The IMT at Nuremberg (November 20, 1945–October 1, 1946), Scott writes, was defined by the chief prosecutor at the trial, US Associate Supreme Court Justice Robert H. Jackson, as "a literal enactment of a judgment of history that had come with victory in war; the war had delivered the verdict, the role of the Tribunal was to put it into effect."[27]

Historians and political theorists have identified the Holocaust and the war tribunals at Nuremberg as the origin of contemporary theories in international law regarding human rights, war crimes, crimes against humanity, genocide, and aggressive war. Nuremberg provided the conceptual foundation for key postwar legal documents that mark the emergence of human rights as we conceive of them today—most notably, the Universal Declaration of Human Rights (1948) and the Geneva Convention (1949)—and establish international standards for the protection of victims of war and violence. It represented the unprecedented prosecution of the political leadership of a sovereign state (Germany) responsible not only for waging aggressive war against other sovereign states, which is now defined as a punishable crime, but also for harms committed against its own citizens and subjects that are now defined as crimes against humanity.[28] Finally, the IMT created a new model for reparations and human rights insofar as individuals and groups who were victims of Nazi terror could for the first time in history make legal claims for compensation and redress against state-sponsored violence and harm.

There is considerable debate among historians and political theorists as to exactly when Holocaust consciousness emerged and when human rights took hold in the popular imagination, a topic I will return to in the conclusion of this chapter. Suffice it to say that however we might date that ascension, the Holocaust today as a universal referent for trauma, for moral and political principles underpinning reparations and human rights, and for the legal assignation of victims and perpetrators remains uncontested. Indeed, the Holocaust has accrued such a singular status that every other modern historical catastrophe is insistently analogized to this master signifier of European disaster. In this regard, we might observe that postwar reparations as a moral response to European violence against Europeans, as well as European reparations for Europeans, rescripts a logic of the self-same subject I explored in the previous chapter. "Trauma theory's failure to give the sufferings of those belonging to non-Western or minority groups due recognition," writes Stef Craps, "sits uneasily with the field's ethical aspirations."[29] I would like to focus on the ways psychoanalysis might be redeployed not only for interrogating the leaps of bad faith that configure the victorious nation-state as the final arbiter of the judgment of history but also for dislodging any one group's privileged hold on traumatized victimhood and repair in a wider history of global violence and genocide in colonial modernity.

Given this context, it is crucial to emphasize the different work that law and psychoanalysis perform—both in tandem and at odds with one another—to align trauma and victimhood with the figure of the injured human being deserving of recognition and reparations. On the one hand, insofar as psychoanalysis is the privileged vocabulary of trauma and therefore dispositive of injury and harm in contemporary discourses of human rights, crimes against humanity, and reparations, law and psychoanalysis work together to produce the privileged figure for rights and representation: the Holocaust survivor. On the other hand, insofar as trauma in psychoanalysis can level distinctions between victims and perpetrators—and insofar as law and the judgment of history are dependent precisely on the nomination of one victim and one perpetrator—psychoanalysis and law are at odds with one another.

This cleaving should not be bemoaned. I offer this observation not to disavow the moral clarity of the Holocaust. Indeed, in the face of resurgent right-wing nationalism across the world today, moral certitude is no doubt something we cannot not want. Rather, I offer this insight as a way to understand more clearly how legal boundaries between victim and perpetrators are secured precisely through the arrogation of trauma, and how these legal boundaries might become displaced and expanded in order to bring justice to those who remain unacknowledged by historical judgment and intent.[30] If trauma functions across psychoanalytic and legal registers to designate particular victims and perpetrators, to produce narratives of cause and effect, to create accounts of innocence and blame, to engender acts of legitimate defense and illegal aggression, to justify the redistribution of property and human life after violent conflict, and to render the judgment of history precisely *under the sign of justice*, then trauma's coming apart in law and psychoanalysis keeps open the possibility, in philosopher Jacques Derrida's words, of "a justice to come."[31] It maintains a space for those left outside the judgment of history to stake a claim—and for those survivors and witnesses such as Tancred to be reattuned to the cries of women, Muslims, Jews, and Christians alike. It creates the potential that their calls for recognition can be heeded.

In shifting from the Holocaust to Hiroshima, from Nuremberg to the Tokyo War Tribunals, and from the space of the Transatlantic to that of the Transpacific, the judgment of history recedes. Unlike Europe, where broad historical consensus on World War I and the Holocaust has been reached, the history of Hiroshima and of war and violence in Transpacific more generally remain a source of bitter acrimony and disagreement to this very day. There is no historical consensus on the confounded relationship between victim and perpetrator in the aftermath of nuclear genocide. How, then, might heightened attention to the history of the traumatized subject illuminate the subject of Cold War history in the Transpacific? Equally so, how does attention to atomic warfare and its Cold War effects work to contest the social and psychic constitution of the Holocaust as universal history?

The Subject of History

On August 6, 1945, the US military detonated Little Boy over Hiroshima. Three days later, they detonated Fat Man over Nagasaki. These two atomic bombs are the only nuclear weapons ever used on a human population. On August 15, 1945, Japan announced its unconditional surrender to the Allied Forces. Government representatives from Allied and Axis powers signed the Japanese Instrument of Surrender aboard the battleship USS *Missouri* in Tokyo Bay on September 2, 1945, thus terminating hostilities in the Transpacific and officially ending World War II. General Douglas MacArthur concluded the event with these solemn words: "Let us pray that peace be now restored to the world, and that God will preserve it always. These proceeding are closed!"[32] MacArthur's wish for peace would hardly come to pass, especially in the embattled space of Asia throughout the ensuing Cold War.

The deployment of Little Boy and Fat Man marked the advent of the Atomic Age in world history, but it also connected the specter of nuclear annihilation indelibly to Asia. Though we imagine atomic destruction today in the language of nuclear universalism—one threatening the existence of every living creature and thing on planet Earth—the Asian origins of "ground zero" must not be forgotten. A little more than one year after the destruction of Hiroshima and Nagasaki, the *New Yorker* provoked a widespread debate on these tangled issues, publishing its most famous volume in the magazine's storied history. The August 31, 1946, issue bore a seemingly innocuous, light-hearted pink and green cover depicting various summer frolics in a bustling and verdant park (fig. 2.1). There was little hint of the disturbing essay that lay inside. That essay was written by John Hersey (1914–93), a World War II correspondent for *Time* and *Life* magazines and a Pulitzer Prize–winning novelist born in Tianjin, China, to US missionaries. It took up the entire magazine, an unprecedented publishing act the *New Yorker* has yet to repeat.

Narrated in restrained and understated prose, "Hiroshima" chronicled the immediate aftermath of the first atomic bombing from the intertwined perspectives of six surviving inhabitants of the decimated Japanese city: a US-trained Methodist minister (the Reverend Mr. Kiyoshi Tanimoto), a white Jesuit missionary from Germany (Father Wilhelm Kleinsorge), a young surgeon working in the city hospital (Dr. Terufumi Sasaki), a retired physician (Dr. Masakazu Fujii), a widowed seamstress with three small children (Mrs. Hatsuyo Nakamura), and a young woman employed as a personnel clerk in a local factory office (Miss Toshiko Sasaki). After spending three weeks in May 1946 interviewing numerous survivors of the event, Hersey had eventually settled on these six civilian witnesses. He asked each of his informants to describe what they had experienced in Hiroshima and to try to make sense of the destruction of what had once been a thriving port city as well as enormous military-industrial complex of 380,000 inhabitants.[33] Each person, Hersey reports, wondered "why they lived when so many others died."[34]

The essay "Hiroshima" started where the "Talk of the Town" column usually begins, and it was accompanied by this concise editorial directive: "The New Yorker this week devotes its entire editorial space to an article on the almost complete obliteration of a city by one atomic bomb, and what happened to the people of that city. It does so in the conviction that few of us have yet comprehended the all but incredible destructive power of this weapon, and that everyone might well take time to consider the terrible implications of its use."[35] William Shawn, the managing editor of the *New Yorker* at the time, believed that despite numerous public debates about, as well as political justifications for, the use of nuclear weapons, insufficient attention had been paid to the "human dimensions" of the atomic bombings. Shawn hoped to bring Japanese voices absent in US accounts of the catastrophe to the American public. Harold

FIGURE 2.1. Cover of the *New Yorker* for August 31, 1946.

Ross, the founder and editor-in-chief of the magazine, supported Shawn's aspirations, noting in a letter to staff writer E. B. White that his managing editor felt the *New Yorker* had a special obligation "to wake people up [to the implications of nuclear warfare], and says we are the people with a chance to do it, and probably the only people that will do it, if it is done."[36]

In scaling up its focus from the "almost complete obliteration" of a Japanese city—approximately seventy thousand people perished instantly, and another fifty thousand died shortly thereafter from radiation poisoning and wounds—to the more expansive goal of awakening "everyone" to the "terrible implications" and "the all but incredible destructive power" of nuclear weapons, the editorial directive attempts to confront the existential threat to human existence initiated by atomic warfare, returning us to McCarthy's epigraph opening this chapter. In so doing, the statement exemplifies how nuclear destruction was transformed into a universal predicament affecting everybody and everything. It simultaneously indexes a psychoanalytic structure of trauma with significant political and ethical implications regarding whose suffering deserves recognition and whose must remain outside consideration. To return to Caruth on the problem of survival and witnessing, Hiroshima is an unclaimed experience that we—the human race—have not "yet comprehended," a violence that has passed us by and one whose significance is yet to be determined. In Caruth's vocabulary, Hersey's reportage draws attention to the problem of how we are implicated in each other's traumas as well as how legal and psychic deployments of traumatized victimhood will come to overdetermine the disaster's historical significance.

The August 31 issue of the *New Yorker* was a historic event in and of itself, and Hersey's article was an instant sensation. In the social media lingo of today, it went viral, rocketing off the newsstands: scalpers demanded fifteen or twenty dollars for a publication whose normal cover price was fifteen cents. Less than two weeks later, beginning on September 9, 1946, the full text of "Hiroshima" was read nationally over four consecutive days on ABC radio, and it was also subsequently broadcast in England, Canada, and Australia.

Harry Scherman, the director of the Book-of-the-Month Club, sent a free copy of "Hiroshima" to his entire membership, remarking that "we find it hard to conceive of anything being written that could be of more importance at this moment to the human race."[37] Hersey's publisher, Alfred A. Knopf, reprinted the 31,000-word article as a short book in October 1946, and it has remained in print ever since, with a circulation of over 3 million copies as well as translations into numerous languages. Variously described as the most celebrated piece of journalism to emerge from World War II and the most significant piece of American reportage in the twentieth century, Hersey's account, as American studies scholar Priscilla Wald observes, "documented the experience not only of the bombing of Hiroshima, but of the bombing as an event that produced a remarkable transformation. Hersey's article retroactively documents the metamorphosis of human beings and of humanity in the wake of Hiroshima."[38]

In part 3 of "Hiroshima," at the dead center of Hersey's four-part essay, the Reverend Mr. Kiyoshi Tanimoto, pastor of the Hiroshima Methodist Church, comes upon a sandspit in Asano Park—one perhaps not so dissimilar from that depicted on the frolicking cover of the magazine—a designated evacuation zone where numerous shell-shocked survivors have gathered on the evening of August 6. There Reverend Tanimoto encounters approximately twenty exhausted men and women, and he beckons them to board the small boat he is piloting. They do not—they cannot—move, and he realizes with sudden alarm that they are too weak to lift themselves away from the rising tide. A modern-day Charon navigating the blazing deltas of the devastated city, Reverend Tanimoto, Hersey writes,

> reached down and took a woman by the hands, but her skin slipped off in huge, glove-like pieces. He was so sickened by this that he had to sit down for a moment. Then he got out into the water and, though a small man, lifted several of the men and women, who were naked, into his boat. Their back and breasts were clammy, and he remembered uneasily what the great burns he had seen during the day had been like: yellow at first, then red and swollen, with the skin sloughed off, and finally, in the evening, suppurated and smelly. With the tide risen, his bamboo pole was now too short and he had to paddle most of the way across with it. On the other side, at a higher spit, he lifted the slimy bodies out and carried them up the slope away from the tide. He had to keep consciously repeating to himself, "These are human beings."[39]

I begin with these entwined emphases on Hersey's "human beings," on Scherman's "human race," and on Ross's "human dimensions" of the atomic bombing because they collectively produced, departing from Wald's compelling insight, a very specific image and transformation of humanity in the irradiated space of postwar Japan.

Remarkably, the figure of the human emerges in Hiroshima *only in the wake of nuclear holocaust*, only in the aftermath of total war and destruction. That is, only after the inhabitants of the devastated city are incinerated and destroyed, only after they become "slimy," "swollen," and "suppurated" creatures, only after they are reduced to nothing more than a repugnant state of blood and viscera, do these former mortal enemies appear to be human at all. What are the psychic and legal implications of this emergence of the human in ruins—this metamorphosis of traumatized Japanese humanity from the ashes and bones of Hiroshima? How do psychic and legal conceptions of victim and perpetrator come together, and how do they fall apart, in these redolent images of suffering, and to what social and political effects?

Psychically, to return to a key point from the first section, this emergence of the human from postcolonial historian Dipesh Chakrabarty's "waiting room of history" depends precisely on the uneven distribution of trauma. If no preexisting human exists in the space of Japan prior to the atomic bombing, Hersey's shocking portrayal of the violated Japanese civilians of Hiroshima transforms them in "a noiseless flash" from inhuman perpetrators of aggressive war into traumatized victims of atomic devastation. Indeed, by drawing attention to the physical ministrations of Reverend Tanimoto and the wretched strangers to whom he selflessly attends, "Hiroshima" not only humanizes the giver and the receiver of care but also raises the question of how such ministrations are dispersed in the aftermath of catastrophe. The recognition of violation and the allocation of care thus produce the figure of the human retroactively, creating in its wake those traumatized subjects deserving of repair.

Equally, by focusing his essay on two Christian clerics—a Japanese reverend trained at Emory University and a white Jesuit missionary from Germany—as well as two Japanese doctors who tend to the sick and wounded, Hersey's article creates a circuit of identification between the violated civilians of Hiroshima and the liberal white readership of the *New Yorker*.[40] Such identifications, as I argued in chapter 1, are the sine qua non of psychic processes of reparation by which the other comes to be recognized as meriting attention and repair. "We are only able to disregard or to some extent sacrifice our own feelings and desires," Klein reminds us, "and thus put the other person's interests and emotions first, if we have the capacity to identify ourselves with the loved one."[41] Ultimately, such identifications created through Hersey's arrogation of traumatized humanity lay the psychic foundation for the emergence of a nuclear universalism binding the West with war-torn Japan and affecting "everybody."

Legally, this postwar redistribution of trauma—the remarkable production of an exhausted and eviscerated Japanese humanity through atomic victimhood—is politically notable given unrelenting Allied depictions throughout World War II of the Japanese enemy as precisely anything but human. Seen as treacherous and fanatic kamikazes all too ready to sacrifice their collective lives in an irrational drive to victory for their emperor, the Japanese, as historian John Dower notes, were "subhuman, inhuman, lesser human, superhuman—all that was lacking in the perception of the Japanese enemy was a human like oneself."[42] It was precisely such dehumanization, Dower observes, that "facilitated the decisions to make civilian populations the targets of concentrated attack, whether by conventional or nuclear weapons."[43] Indeed, an official US intelligence review in July 1945 on the deployment of strategic air power in Japan reads (in all capital letters), "THERE ARE NO CIVILIANS IN JAPAN."[44]

While "Hiroshima" begins to enact this metamorphosis psychically in August 1946 through provocative representations of a devastated humanity, the project takes on increasingly urgent and expanded legal form throughout the postwar occupation of Japan by US-led Allied Forces. How is it that Germany is rendered eternal perpetrator through the judgment of history, even as Japan emerges as traumatized victim deserving of repair outside any such historical judgment? If, as I emphasized above, psychic and legal appropriations of trauma work together to maintain a strict boundary between victim and perpetrator, how is it that a vanquished Japan and a defeated Germany end up on opposite sides of this binary?

In the closing pages of "Hiroshima," Hersey observes that many citizens of the destroyed city "continued to feel a hatred for Americans which nothing could possibly erase. 'I see,' Dr. Sasaki once said, 'that they are holding a trial for war criminals in Tokyo just now. I think that they ought to try the men who decided to use the bomb and they should hang them all.'"[45] Suffice it to say, a war tribunal against the American victors for their deployment of not just one but two atomic weapons on civilian populations in Japan in what Dower describes as a "war without mercy" was as unthinkable in 1946 as it is today.[46] If Japan emerges as victim of atomic disaster, there can be no implicated perpetrator of this violence. To return to an insight from Scott, a victorious US nation-state fighting "the good war," in Studs Terkel's terms, can only be the adjudicator, not the violator, of crimes against humanity. In this regard, we might speculate that the emergence of traumatized humanity in Hiroshima—indeed, the remarkable transformation of Japan from evil object into human subject—serves as a compensatory mechanism meant not to indict the US nation-state but, rather, to ensure its sovereign innocence and, in turn, that of Japan. What are the psychic and legal logics of this gesture?

In stark contrast to the Holocaust in Germany, the deployment of nuclear weapons in Japan has never been legally categorized as a "crime against humanity" or as "genocide." Among weapons of mass destruction—nuclear, chemical, biological—only nuclear weapons are not prohibited in international law. This arrangement, Antony T. Anghie observes, highlights the fact that "sovereignty *is* nuclear weapons."[47] Dr. Sasaki's impassioned outburst lamenting the total destruction of his hometown thus raises an unresolved problem of historical response and responsibility marking the dream of human rights. It invokes, that is, the subrogation of the sovereignty of the victorious nation-state in the service of protecting the sovereignty, the sanctity, of the eviscerated individual. This aborted dream comes to frame the entire legal proceedings of the IMTFE, with significant consequences for those killed who remain eccentric to the judgment of history.

The IMFTE, or Tokyo War Crimes Tribunal, convened on May 3, 1946, and adjourned over two and half years later, on November 12, 1948. It was modeled on the IMT at Nuremberg, whose earlier prosecutions of Nazi war criminals served as the legal precedent for proceedings in Tokyo. Comparatively streamlined, Nuremberg commenced on November 20, 1945, and concluded less than a year later, on October 1, 1946. Nuremberg exemplified an unprecedented model of international cooperation as Allied forces, led by prosecutors from the United States, sought to account for atrocities committed by the top leadership of the Third Reich. At the same time, it developed a host of new categories referenced above for prosecution in international law through what Schmitt lamented as the increasing criminalization of war in the first half of twentieth-century Europe.[48] The IMT was overseen by judges from four Allied Powers: France, Great Britain, the Soviet Union, and the United States. It effectively produced unanimous judgments on the German leaders and organizations charged before it, providing a "dream of deliverance"—an image of Nazis as eternal perpetrators and Jews as eternal victims in the popular imagination.[49]

In contrast, this final judgment of history could not have been further from what was achieved in Tokyo. Following Japan's surrender, General MacArthur became the Supreme Commander for the Allied Powers (SCAP) who led the postwar occupation and reintegration of Japan into the (Western) family of nations. Although he was initially appalled by the various demands of the defeated Japanese Empire for relief, it is a "great irony of the subsequent history," as legal historian Harry N. Scheiber observes, "that in MacArthur's oversight of Japan in the Occupation era, he became the controlling figure in a process that in fact did work with great effectiveness to 'relieve Japan of . . . the physical and psychological burdens of defeat.'"[50]

In its efforts to suppress the spread of communism in Asia through the rollback and containment of Soviet and Chinese insurgency, and in its determination to transform what had been long histories of European domination through direct colonialism and extraction in the region into structures of political modernization and economic development, the United States came quickly to prioritize the establishment of a postwar US-Japan Cold War political alliance. To the extent that the atomic bombings documented by Hersey began to transform the enemy nation psychically from being perpetrators of inhuman violence to victims of nuclear catastrophe for his Western audience, the IMTFE and the seven-year Allied occupation of Japan also legally shifted the country's own self-image from that of colonizer to that of colonized.

Here the psychic and legal worked in tandem to produce in the social imagination a traumatized Japanese humanity largely innocent of any war crimes—indeed, to transform prewar Japan from enemy and stranger into postwar ally and friend of the West. As a consequence of this transformation, the inter-Asian studies scholar Chen Kuan-hsing observes, Japan was not forced to engage in "the reflexive work of deimperialization within its own territory and . . . grappling with its historical relations with its former colonies (Korea, Taiwan, and others) or its protectorate (Manchukuo)."[51] To be sure, from the perspective of their prior colonial subjects, Japan is hardly innocent. The IMTFE and its legacy, as transnational feminist scholar Lisa Yoneyama elaborates, "might as well be regarded as a showcase not so much for 'Victor's Justice' as for 'Victor's Exoneration,' . . . [a] Cold War culture of impunity under U.S. regional hegemony."[52]

At MacArthur's insistence, Emperor Hirohito, the wartime head of the Japanese nation-state, was granted special immunity. Unlike the top political leadership of Germany, Hirohito was not subject to trial and was spared even from testifying at the IMTFE. MacArthur believed that protecting Hirohito and permitting the emperor system to continue into the postwar period was critical for stabilizing Japan politically and rehabilitating its economy for capitalist development as one critical bulwark against the spread of communism in Asia. Although the Japanese Constitution was overhauled and many reforms were introduced to secularize Hirohito's "divine" powers and to limit future militarization, Hirohito was not deemed a war criminal and was allowed to remain on the Chrysanthemum Throne. As Dower writes, "With the full support of MacArthur's headquarters, the [IMTFE] prosecution functioned, in effect, as a defense team for the emperor."[53] Remarkably, unlike with Nuremberg, the defense team for the Japanese leadership on trial in Tokyo was, in part, staffed by US lawyers.

In Hirohito's stead, select high-ranking military officials were prosecuted and held responsible for leading the Japanese people and the emperor himself astray. Hirohito was characterized as a victim of circumstance, a naive and innocent ruler manipulated by a cadre of fanatical military leaders led by General Hideki Tōjō. In short, as legal scholars Yuki Tanaka, Tim McCormack, and Gerry Simpson observe, "the Emperor, too," like the civilian populace, "was a victim of the war."[54] This image of traumatized innocence, as historian Alexis

Dudden argues, would adhere to "how the Japanese people in general would eventually be described in the preponderance of national storytelling that took its cue from this decision."[55]

Astonishingly, in March 1948, three years after the conclusion of World War II, MacArthur insisted on a suspension of any reparations policy "not only on economic grounds but also as a matter of justice because 'Japan has already paid over fifty billion dollars by virtue of her lost properties in Manchuria, Korea, North China and the outer islands.'"[56] Here MacArthur's stance replicates the colonial logic of Locke and Indigenous dispossession analyzed in chapter 1. The general describes Japan as the victim of property theft, as if these occupied territories were Japan's property in the first instance. In Locke's terminology, MacArthur constitutes Japan as an "adversary" rather than "aggressor"—as victims defending themselves from the postcolonial populations they once ruthlessly colonized, conquered, and governed across East and Southeast Asia prior to their defeat in the war. Even more, through this reversal MacArthur implicitly shifts the responsibility for Japan's victimhood from atomic devastation wreaked by US and Allied forces to the nation-state's former colonial subjects and their theft of Japan's lands.

The 1951 San Francisco Peace Treaty normalizing relations between Japan and the Allied Powers formalized this astounding and compromised position on Japan's responsibilities for redress.[57] Subsequent treaties of normalization between Japan and the United States, Korea, Taiwan, and China followed suit, prohibiting individuals and groups harmed by Japanese militarism and colonialism from filing legal grievances for reparations. Flying in the face of growing postwar movements for decolonization across the Third World, MacArthur's stance thus aligned Japan—rather than its former colonial subjects—with Locke's aggrieved colonial settlers defending themselves in the New World.

Japan was thus relieved, as Scheiber argues, of both the material and the psychological burdens of defeat—of response and responsibility—for military officials and civilians alike. The revival of Japan's war-torn economy, as well as its restoration to the (Western) family of nations through its suturing into a new international economic order, forestalled, and continues to forestall, any judgment of history or restoration of Japan's violated colonial subjects. This suspension of history is mobilized through the emergent logics of reparations and the human framed by Hersey's seminal reportage, enforced by MacArthur's occupation policies producing an image of Japanese innocence and victimhood, and internalized by Japanese leaders and citizens alike.

We thus witness how psychic and legal appropriations of trauma worked to sustain an image of Japan that served to prevent any reckoning of the country with its prewar history of colonization and violence, as well as their postwar effects, throughout the Greater East Asia Co-prosperity Zone. This history of the traumatized Japanese subject returns us to the enduring problem of sovereignty and the modern racial state. A reformulated US-Japan partnership in full bloom within only half a dozen years after the end of an unspeakably inhuman war might be seen as quite remarkable.

Certainly, this alliance testifies to how perceived geopolitical mandates could radically reshape racial attitudes toward the Japanese in the postwar period, producing them as human beings worthy of recognition and rescue while laying the foundations for the emergence on the other side of the Transpacific of (Japanese American) model minority citizenship—what US legal historian Mary Dudziak has described as the international and domestic coproduction of "Cold War civil rights." In effect, as the prewar Japanese colonial empire was succeeded by postwar US military might through a series of brutal wars and partitions, US empire was grafted onto Japanese colonialism, with grave consequences for a much longer history of unacknowledged violence across the region. From this perspective, US empire is not dissociated from prewar histories of Japanese domination in the Transpacific; rather, it converges with them.

Unlike the IMT, the IMTFE consisted of judges and prosecutors from eleven Western Allied powers (Australia, Canada, France, Holland, New Zealand, the Soviet Union, the United Kingdom, and the United States), in addition to China, India, and the Philippines. The participation of these Asian countries made the Tokyo War Crimes Tribunal as much a multiracial as a multinational event, yet little commentary on the IMTFE's racial implications has been proffered by scholars in either international law or Asian studies. However, judges and prosecutors from Taiwan, Korea, Malaysia, Singapore, Indonesia, Burma, and Indochina—colonial protectorates and other sites of Japanese occupation from the late nineteenth century—were notably absent from the Tokyo War Crimes Tribunal.

Their omission, as historian Yuma Totani notes, led to accusations of racial bias, indexing the enduring problem of race, racism, racial reparations, and the inhuman in colonial Asia.[58] By focusing on war crimes against white prisoners of war and civilians, the IMTFE ignored numerous aggressions committed by the Japanese Imperial Army against its Asian colonial subjects. According to Totani, Allied powers, particularly the United States, withheld evidence of certain sensitive war crimes cases, including the "'comfort woman' system, medical experimentation and the bacteriological warfare committed by Unit 731, and atrocities targeted at the Asian civilian populations" deemed less than human.[59]

In other words, the IMTFE focused on war crimes committed by colonizers (Japan) against other colonizers (Western Allied powers) rather than war crimes committed by colonizers against the colonized—a closed circuit of injury and repair predicated on the universalizing project of European humanism.[60] After all, what a colonizing Europe did to the world, a colonizing Japan did to Asia, earning in the process its exemplary status as the only "modern" nation-state in the region. As historian Takashi Fujitani observes in *Race for Empire*, Japan and the United States were not just mortal enemies in World War II but also parallel empires constructed on notions of Japanese and white racial supremacy, respectively, in their projects of total war. For instance, as the Japanese conscripted colonized Korean male subjects into their Imperial Army, racialized Japanese American males were coerced into enlisting in the US military as a way to prove their loyalty to the US nation-state and to secure release from the concentration camps where they and their families were detained en masse throughout World War II. In the process, to refigure McCarthy's trenchant observations, Hersey's "insipid falsification of the truth of atomic warfare" raise the urgent question of exactly which murdered and sacrificed populations Hersey would have had to interview in order to bring justice to the dead.

Unsurprisingly, then, the verdicts reached at the IMTFE were riven with disagreement and dissent. Of the eleven justices, eight delivered a majority opinion of guilt, qualified by two concurrent opinions, and opposed by three separate dissenting opinions.[61] In contrast to Nuremberg, there would be no final judgment of history in Tokyo. There was not even, as historian Marc Gallicchio observes, agreement on what to call the war.[62] In the final analysis, without any consensual postwar history regarding Japanese empire in the Transpacific in the aftermath of atomic war, any prospects of making legal claims on the defeated nation-state for injury and redress on the part of its prewar colonial subjects were rendered moot. From this perspective, a consensual history becomes the sine qua non for any possibility of racial reparations and the human.

In a 1,235-page opinion dissenting from each and every guilty verdict handed down by the tribunal, Justice Radhabinod B. Pal of India, a forthright critic of Western colonialism, insisted that a judgment on Japan's war crimes could not be made in good conscience without first giving an account of prior atrocities committed by Western powers in Asia and Africa. To do otherwise would be legally untenable.[63] "Was it not the West that had coined the word 'protectorate' as a euphemism for 'annexation'?" Pal asks. "And has not this constitutional fiction served its Western inventors in good stead?"[64] Pal argued that the Japanese Kwantung Army and Western Allies were partners in crime. He condemned Western powers for prosecuting Japan while ignoring their own atrocities rooted in much longer colonial histories of Western dominance leading up to the atomic bombings.[65] Indeed, Pal proffered the unthinkable analogy between genocide in Europe and nuclear holocaust in Asia, directly connecting the Final Solution of a defeated Germany with the deployment of nuclear weapons on Hiroshima and Nagasaki by a victorious United States.[66]

This, of course, is the precise analogy that the racial state and its judgment of history refuses to, and cannot, confront. However, it is an analogy that is exposed precisely by the (mis)alignments between legal and psychic theories of the history of the traumatized subject. As holocaust and genocide migrate from Europe to Asia, the coming together of psychoanalysis and law in a history of trauma circumscribes a Western limit to the figure of the human—to crimes against humanity and to war crimes—and restricts their legal and ethical ambit. This history of the traumatized subject thus underwrites the subject of an unresolved colonial history of violence, one beginning with European exploration and expansion, extending to Japanese empire-building throughout East and Southeast Asia, and inherited by an ascendant US nation-state during the American Century.

To be sure, as the comparison between the Holocaust and Hiroshima emphasizes—and as transitional justice movements across the decolonizing world such as South Africa's Truth and Reconciliation Commission underscore—when the sovereign integrity of the victorious nation-state is deemed to be the political priority, a juridical process of reckoning with violence and conflict will insistently recognize a proliferating chain of traumatized victims in the absence of clear perpetrators held to legal accountability. As a result, the psychic and legal arrogation of trauma come together not to uphold a strict boundary between designated victims and perpetrators but, rather, to generate a politics of liberal recognition—a politics of individual suffering and identity—that bypasses the judgment of history, an ethics of relationality and responsibility, the imperative for material redistribution and redress, and the will to justice. If a politics of victim and perpetrator produced through the consensual judgment of history is the sine qua non of any legal claim for reparation—as the unresolved record of comfort women in the Transpacific underscores—we ought to begin theorizing how to reconfigure repair and redress outside paradigms of sovereignty altogether. We must do so neither for the sake of the victorious nation-state nor in the name of the human but on behalf of those rendered inhuman by their fraught political legacies and loaded psychic dynamics.

Human-Civil-Inhuman

How, then, do we recognize and represent the cries of the dead and unheeded? My conclusion moves from the realm of history and law to the domain of literature and psychoanalysis to reconsider the legal imperative to nominate one victim and one perpetrator—to give a sanctioned historical account of cause and effect, to assign a moral calculus of blame and blamelessness, and to proffer a judgment of guilt and innocence—in the name of the human and human rights. Describing Freud's turn to Tasso's poem, Caruth writes, "If Freud turns to literature to describe traumatic experience, it is because literature, like psychoanalysis, is interested in the complex relation between knowing and not knowing."[67]

In this concluding section of chapter 2, I start from Caruth's observations on a nonreferential theory of traumatic history, along with the psychoanalytic insight that trauma emerges from a diagnostic realm that remains beyond judgments of good and evil, to explore how a crisis of witnessing might be refigured for a justice to come. Butler notes, "Fictional narratives in general require no referent to work as narrative, and we might say that the irrecoverability and foreclosure of the referent is the very condition of possibility for an account of myself."[68] In turn, such an account provides the possibility by which moral judgments might be suspended so that an "ethical reflection on the humanity of the other, even when that other has sought to annihilate humanity," might be able to emerge.[69]

The historical fiction of Kazuo Ishiguro and Chang-rae Lee exploring the aftermath of war and violence in the Transpacific allows us to examine the relationship between moral judgment and ethical reflection in the face of knowing and unknowing—a relationship producing social and political categories of the human, inhuman, and the civil. The atomic bombings of Japan established a doctrine of preemptive violence in Asia that would come to define conflict in the region throughout the Cold War period. The rollback and containment of communism in the name of freedom justified the US occupation of various Asian countries, which became the staging ground for its interests, the exhibition of its military force, and the display of its technological mastery as an ascendant US nation-state sought to secure the ideological and material conditions for the global circulation of capital in the postwar era. Preemptive violence marked the socialization of the death drive in Asia—a repetition compulsion of destruction that commenced with the atomic devastation of Hiroshima and Nagasaki, moved across a decimated Korean peninsula, and continued to the napalmed jungles of a ravaged Vietnam. In *Cold War Ruins*, Yoneyama notes that preemptive violence under US hegemony left numerous colonial legacies in the Cold War Transpacific intact, thereby sustaining an inability to redress certain violences.[70] As the historical referents of these events are rendered illegible by government policies and military action, and as the possibilities of historical consensus and legal judgment become suspended, literature offers the prospect of imagining otherwise by rethinking the politics and ethics of repair and responsibility under erasure.

Ishiguro, who was born in postwar Nagasaki in 1954 and immigrated to the United Kingdom at age six, explores the psychic production of Japanese innocence and bad faith in the wake of nuclear war and the Allied occupation of Japan. Lee, who was born in Seoul in 1965 and immigrated to the United States at age three, conducts a similar investigation in relation to American exceptionalist discourses of civil rights that seek to bury unresolved histories of Western empire in the Transpacific and, in particular, the Japanese colonization of Korea. Bringing together Ishiguro and Lee, I consider how the uneven distribution of the human in postwar Japan, the redress for the wartime incarceration of Japanese Americans in the name of civil rights, and the largely unsuccessful legal claims for reparations by comfort women rendered inhuman by the Japanese Imperial Army collectively rework conventional distinctions between victims and perpetrators in order to offer my own account of a nonreferential theory of traumatic history and repair in the Transpacific.

As discussed above, the decision not to prosecute Emperor Hirohito, who was himself characterized as a victim of the war machine led by his generals, created a postwar image of traumatized innocence, which affected how Japanese civilians would come to see themselves in the "preponderance of national storytelling that took its cue from this decision."[71] Ishiguro's early novels might be described as collectively delineating the construction of everyday Japanese subjectivity in relation to these political mandates—a subjectivity relieved of the material and psychological burdens of defeat. Far removed from the time and space of the battlefield or from official judicial proceedings meant to reckon with the grand histories of war and its towering agents of destruction, Ishiguro's protagonists for the most part are ordinary if not self-justifying narrators who are plagued by solipsism and unmetabolized bad faith. In somewhat different terms, Ishiguro transmutes grand historical landscapes of collective war and violence into individual crises of trauma and witnessing that index an unacknowledged complicity in the suffering of others. As with Tancred, the struggles of his protagonists mark a disavowed history of violence and responsibility that often passes them by, even as their struggles symptomize an inadvertent form of displaced testimony and self-indictment. An unsettling history of violence is transformed into a history of the unsettled subject.

Take, for example, the loyal butler, Stevens, in Ishiguro's third novel, *The Remains of the Day* (1989), who reflects on his actions in the UK countryside during the interwar years. The entire narrative traces Stevens's ruminations on "his small contributions to the course of history"—that is, what it meant for him to have served in the household of his former employer, Lord Darlington, an exalted British aristocrat who helped to facilitate not only the political rehabilitation of Germany following World War I but also the rise of German fascism under the Third Reich as an unabashed Nazi sympathizer. Ishiguro sets his debut novel, *A Pale View of Hills* (1982), in the atomic ruins of Nagasaki and focuses on the maternal struggles between two mother-daughter pairs as they navigate their conflicted desires to immigrate on the coattails of pale partners to the United States and United Kingdom in the wake of the bombings and Allied occupation. We encounter in Ishiguro's first novel a minor character similar to Stevens, the retired teacher Ogata-San. Due to his unrepentant nationalism throughout and after the war, Ogata-San's reputation in the community has been tarnished. In a brief exchange with his adult son Jiro, one directly referencing textbook-related controversies about Japan's checkered war history that continue to this day, Ogata-San laments that under US occupation the Japanese education system has been decimated:

> "I devoted my life to the teaching of the young. And then I watch the Americans tear it all down. Quite extraordinary what goes on in schools now, the way children are taught to behave. Extraordinary. And so much just isn't taught any more. Do you know, children leave school today knowing nothing about the history of their own country?"
>
> "That may be a pity, admittedly. But then I remember some odd things from my schooldays. I remember being taught all about how Japan was created by the gods, for instance. How we as a nation were divine and supreme. We had to memorize the text book word for word. Some things aren't such a loss, perhaps."[72]

This exchange between father and son—in particular, Ogata-San's observation that "children leave school today knowing nothing about the history of their own country"—is especially ironic, given the nation's unresolved history with its colonial past as well as its self-image of victimhood. Ogata-San and Jiro's exchange captures the predicament of Japan's unsettled relationship to its "divine" origins—its imperial destiny and strident militarism—even, and especially, as this history of violence is indemnified and rewritten by US occupiers under the secular signs of capitalism and liberal democracy.

In this particular example, Ishiguro presents the conflict as an intersubjective struggle between father and son, between an older and a younger generation, and between the perspectives of a traditional Japan and secular West. As we soon come to witness, this debate is not just staged *between* contending subjects as a conflict of social value and generational difference. More accurately, it comes to inhere *within* every character we encounter throughout Ishiguro's early novels on war as an everyday *intrasubjective* psychic condition of bad faith and conscience.

In his second novel, *An Artist of the Floating World* (1986), Ishiguro provides a sustained investigation of the social construction and political effects of this intrasubjective bad faith, presenting an expanded study of Ogata-San in the character of Masuji Ono, a retired painter in occupied Japan. Ono struggles throughout the novel to reconcile Japan's prewar and postwar histories with his own. Over the course of three chapters spanning from October 1948 to April 1949 to November 1949, and overshadowed by the actual trials of the IMTFE, we are told from Ono's limited first-person perspective that during World War II the artist had employed his considerable talents for right-wing nationalist propaganda. Ono produced political art for the war effort while also volunteering as a police informant for the Committee of Unpatriotic Activities, monitoring the artistic productions of his peers and protégés alike.[73]

In the aftermath of Japan's defeat and occupation, Ono's reputation has suffered, and he tells us that he has been discredited for his wartime efforts. Like many of Ishiguro's other protagonists, Ono repeatedly justifies his behavior by telling us that he "acted in good faith," that he believed in all sincerity he was "achieving good for [his] fellow countrymen," and, with repeated echoes to the butler Stevens, that at least he had the "courage and will to try" where others had failed.[74] However, as the novel progresses, the artist begins to develop incipient pangs of regret, a trickle of misgivings that perhaps he may be more culpable than he can readily admit, and that he may have made some errors in the past. The artist begins to recognize that his wartime activities may have caused others to suffer needlessly.

What drives this nascent self-awareness and nagging self-doubt is neither Ono's diminished reputation nor a sense of guilt and responsibility toward untold others but rather a crisis in kinship: Ono's younger daughter, the twenty-six-year-old Noriko, remains unwed. Noriko and her older sister Setsuko both suspect that their father's dubious actions during the war may be negatively influencing Noriko's prospects of securing a suitable marriage proposal. Indeed, negotiations for her marriage a year earlier to one Jiro Miyake were unexpectedly terminated by the groom's family without proper explanation. It is this kinship trouble that compels Ono's self-reflections and instigates a reevaluation of his wartime behavior.

At a marriage negotiations dinner with a new prospective groom, Taro Saito, and his family, Ono offers an unexpected public confession. Toward the end of a disastrous meal in which a typically vivacious Noriko appears awkwardly subdued, Taro's parents raise the name of Mr. Kuroda, a mutual acquaintance and an ex-protégé on whom Ono had informed during the war. In response, the artist makes an announcement to his assembled guests:

> "There are some, Mrs. Saito," I said, perhaps a little loudly, "who believe my career to have been a negative influence. An influence now best erased and forgotten. I am not unaware of this viewpoint. Mr. Kuroda, I think, is one of those who would hold it."
>
> "Is that so?" Perhaps I was mistaken about this, but I thought Dr. Saito was watching me rather like a teacher waiting for a pupil to go on with a lesson he has learnt by heart.
>
> "Indeed. And as for myself, I am now quite prepared to accept the validity of such an opinion."
>
> "I'm sure you're being unfair on yourself, Mr. Ono," Taro Saito began to say, but I quickly went on:
>
> "There are some who would say it is people like myself who are responsible for the terrible things that happen to this nation of ours. As far as I'm concerned, I freely admit I made many mistakes. I accept that much of what I did was ultimately harmful to our nation, that mine was part of an influence that resulted in untold suffering for our people. I admit this. You see, Dr. Saito, I admit this quite readily."[75]

Ono's speech is astonishing for a number of reasons. Most immediately, it illustrates the ways that a history of Japanese militarism throughout Asia is configured as a closed circuit of harm, as an injury to the self-same subject. That is, Ono "freely admits" that his actions were "ultimately harmful to *our* nation, that mine was part of an influence that resulted in untold suffering for *our* people." Nowhere in this confession is there an acknowledgment that his "many mistakes" might have harmed untold *others*—that the "terrible things" for which Ono is responsible might have injured others outside the "divine" island nation-state, those colonized by Japan.

Ono's speech is a literary illustration of Klein's colonial object relations in the context of Cold War Asia. It exemplifies the contemporary psychic production of colonial superego: a closed circuit of injury, repair, and judgment by which the Japanese people are configured as both subject and object, as both perpetrators of violence and victim of harm deserving repair. As with Klein's explorer in the New World, the "wished-for restoration" is given not to the Native populations who suffered "ruthless cruelty" at the hand of colonial settlers who "not only explored, but conquered and colonized"; rather, it is bestowed on brothers and sisters of the same Japanese tribe, to "people of their own nationality."[76]

Here Ishiguro traces the ongoing evolution of liberal guilt and negative narcissism in postwar Japan, the production of a resplendent bad faith aimed at relieving the psychic burdens of guilt and responsibility threatening the national integrity of the Japanese ego. If Ono's speech is "a lesson he has learnt by heart," it also a psychic refraction of a political pedagogy formed by the postwar US occupation that he has internalized, a pedagogy that works to create an unrepentant Japanese subject. In short, as Japanese atomic injury takes priority over the guilt of atrocities committed against countless colonized others, Ono's arrogation of victimhood obviates histories of violence in Japan's imperial empire.

In this regard, I would like to underscore that the crisis in kinship propelling Ono's dubious account of himself is ultimately a crisis of national kinship: an imperative to propagate the Japanese family line. In Klein's pregnant terms, what triggers Ono's incipient sense of misplaced guilt in the aftermath of nuclear destruction is a national mandate of "repopulating the country with people of their own nationality." Ono's guilt psychically manages a history of colonial violence precisely by displacing it, by colonizing the suffering of others for himself. If artists are often said to exploit the suffering of others for their art, Ono's bad faith is here directed instead toward exploiting such suffering for the aggrandizement of the nation-state, a negative narcissism for the self and for the motherland.

Configured less as a political conflict between Japan and the United States, once enemies and now allies, or even an intergenerational rift between parents and children, Ono's speech illustrates how the postwar Japanese subject psychically refracts Cold War political dynamics to produce a history of everyday bad faith and subjective consciousness in the postwar period. In turn, these psychic dynamics and their social effects construct a collective national amnesia, an eschewing of any ethical recognition and responsibility toward the suffering of Japan's colonized populations. Unsurprisingly, even as he offers this dubious confession, Ono's performative utterance comes to undo itself.

In the closing pages of the novel, Ono's entire account of his wartime actions is put into question. In a final conversation with Setsuko, Ono expresses great satisfaction regarding Noriko's upcoming marriage to Taro Saito. He acknowledges his daughter's warnings to take "precautionary steps" concerning his tainted past, but Setsuko disputes her father's version of the dinner events:

> "Noriko told me she was extremely puzzled by Father's behavior that night. It seems the Saitos were equally puzzled. No one was at all sure what Father meant by it all. Indeed, Suichi [Setsuko's husband] also expressed his bewilderment when I read him Noriko's letter."
>
> "But this is extraordinary," I said, laughing. "Why, Setsuko, it was you yourself who pushed me to it last year. It was you who suggested I take 'precautionary steps' so that we didn't slip up with the Saitos as we did with the Miyakes. Do you not remember?"
>
> "No doubt I am being most forgetful, but I am afraid I have no recollection of what Father refers to."[77]

By the end of *An Artist of the Floating World*, the problem of historical response and responsibility, as well as individual agency and witnessing, becomes a problem *writ large* for everyone. Setsuko has "no recollection of what Father refers to," suggesting that the history of violence to which Ono confesses is entirely a fabrication of the old man's imagination. "Forgive me, but it is perhaps important to see things in a proper perspective," Setsuko admonishes him. "Father painted some splendid pictures, and was no doubt most influential amongst other such painters. But Father's work had hardly to do with these larger matters of which we are speaking. Father was simply a painter. He must stop believing he has done some great wrong."[78]

A circuit of intergenerational psychic repression and bad faith is complete, as a blanket of historical amnesia, receding agency, and "no doubt" comes to envelope Ishiguro's second novel. This lack of consensual history comes to bind the older and younger generations in postwar Japan, a crisis of witnessing that blurs the boundaries between victims and perpetrators and, in the process, absolves both Ono and Japan of having done "some great wrong." At the conclusion of *An Artist of the Floating World*, historical consensus recedes, replaced by an intrasubjective as well as intersubjective fantasy of national innocence and enclosure that, indeed, forecloses the possibility of giving an account of oneself. This blurring of the boundaries between victims and perpetrators—indeed, this arrogation of both positions by the postwar Japanese subject—persists as a haunting and unresolved ethical quandary of witnessing throughout Ishiguro's oeuvre.[79]

Dudden observes that from the mid-1950s until the early 1990s, "Germans deliberated their nation's attempted annihilation of the Jews but not the firebombing of Dresden, whereas Japanese ruminated on the wastelands of Hiroshima and Tokyo at the cost of confronting Japan's devastation of large parts of populations of Asia."[80] If the atomic bombings and the Tokyo War Tribunals relieved both the Japanese leadership and, in turn, the Japanese people from politically confronting their past, producing Japan's endemic historical amnesia, Ishiguro traces the elaborate psychic production of this vanishing agency and disavowal of responsibility at the heart of the country's postwar renewal. The inability of either the IMTFE or the Japanese populace to reckon with atrocities committed against tens of millions of colonial subjects in Japan's prewar empire attests, as Yoneyama notes, "to the elisions and exclusions that have underwritten the genealogy of the West-centric notion of 'humanity.'"[81] These elisions and exclusions find their social and psychic template, as I explored in chapter 1, in New World discovery and dispossession—and in Japan's appropriation of the colonial logic of Locke's thief and of Klein's restoration of mother as motherland. In the same breath, they also frame the problem of racial reparations and the human in a string of unresolved conflicts across the twentieth-century Transpacific.

Moving east, Chang-rae Lee extends Ishiguro's reflections on postwar Japan in relation to a declining Europe to consider more systematically how an ascendant US nation-state instrumentalizes the figure of the Asian American model minority to manage unresolved histories of violence connecting Japanese and Western empires in Asia. Moreover, in collapsing the political and psychic distance between victims and perpetrators, and between aggression and repair, Lee illustrates how these histories become entangled in—as well as occluded by—figurations of the human, the civil, and the inhuman.

Lee's second novel, *A Gesture Life* (1999), is narrated from the perspective of Franklin Hata, an immigrant from Japan and the owner of a small medical supplies company in Bedley Run, a leafy fictional suburb in Westchester County, New York. Hata possesses all the stereotypical trappings of the American Dream: a handsome Tudor in an upscale neighborhood, a late-model Mercedes, a sizable nest egg for his impending retirement. By all appearances, he is an exemplary racial subject, embodying ideals of upward mobility and immigrant success defining the postwar rehabilitation and transformation of Japanese aliens and enemies—many of whom were incarcerated by the US state during World War II—into Asian American model minority citizens and friends. Well regarded by his Bedley Run colleagues and neighbors alike, Franklin is known in the community simply as "Doc Hata."

This placid image of Asian American model minority success is, however, steadily eroded over the course of Hata's vertiginous narrative. As it turns out, Doc Hata is neither a doctor nor Japanese, although he is passing for both. Hata is, in fact, a Zainichi, a Japanese colonial subject of Korean descent, part of an outcast racial group forcibly brought to Japan as conscripted labor following the country's occupation of Korea in the late nineteenth century.[82] Adopted from his Korean ghetto of hide tanners and renderers by a childless Japanese couple, Hata—shortened from Kurohata, meaning "black flag" in Japanese—describes this moment as "the true beginning of 'my life[,]' . . . when I first appreciated the comforts of real personhood, and its attendant secrets, among which is the harmonious relationship between a self and his society."[83]

Rendered human through his assimilation into what he describes as Japan's "purposeful society," and exceedingly grateful for the "comforts of real personhood" that this status provides, Hata fights in World War II as an ersatz national subject for the glory of the Japanese Empire. To employ Klein's vocabulary, Hata is simultaneously a good and a bad object. Whether as ersatz Japanese national as palimpsest for degraded Korean Zainichi, or as Asian American citizen as palimpsest for the brutal Japanese enemy, Hata might be characterized as at once valued and denigrated, victim and perpetrator. His condition embodies the social and psychic paradox of Asian American model minority subjectivity all too well.

During the war, Hata serves as a medic in the Japanese Imperial Army, where he is stationed in Burma and attends to a group of enslaved Korean comfort women. Hata forms a bond with one particular "volunteer" known as K, with whom he falls tragically in love—though we might question exactly what love means under such obscene conditions of violence and duress. As the literary scholar Anne Cheng observes in her reading of Lee's novel, "Hata's words fold rape into love, strain into promise, highlighting the disturbing gap *and* intimacy between rhetoric and reality, between romance and coercion. The universalizing language of romance here authorizes forms of violence and domination."[84]

Ultimately, we discover over the course of his unraveling narrative, Hata is in large part responsible for K's demise. He bears witness, like Tancred, to his beloved's destruction, but with a salient difference: he slays her not directly but indirectly. Hata's *lack* of action in response to K's repeated pleas—either as a dehumanized fellow Korean or as the object of his desire and devotion—to end her misery results in an even more gruesome fate. In the end, she is gang-raped and murdered by a band of fellow soldiers in Hata's regiment, her pregnant body dismembered and scattered in the forest. For Hata, saving and killing K are disturbingly one and the same.

We are thus brought back to Tancred's dilemma of survival and witnessing—to the question of a traumatic history of violence that passes Hata by at moment of its occurrence. Considered in light of my discussion of Klein, as acts of love and hate, deeds of care and cruelty, and moments of aggression and repair in *A Gesture Life* become virtually impossible to distinguish from one another, reparation is configured not as the antidote to nationalist violence but, rather, as its animating source. Similarly, as the boundaries dividing Japanese from Koreans, humans from inhumans, colonizer from colonized, and victims from perpetrators erode in the novel, we are left to wonder what exactly is at stake in Lee's deconstruction of these various boundaries. Hata is thus faced with the question of how to live and to take responsibility as *both* the victim and the perpetrator of a colonial history that overdetermines him. In this regard, how do we understand the ethical consequences of Hata's extreme assimilation and passing on both sides of the Transpacific—first as patriotic Japanese subject and then as compliant Asian American model minority citizen—in terms of the politics of reparation?

In Caruth's account of *La Gerusalemme liberata*, it is not until the repetition of violence—Tancred's second killing of Clorinda in the enchanted forest—that the hero is confronted with the ethical dilemma of his actions and with the quandary of response and responsibility. In Hata's case, however, this repetition and confrontation is played out in the forests of Burma. It is enacted across the space of the Transpacific and in terms of Hata's relationships with two overlapping bodies: that of the deceased K in Asia and the living Sunny, Hata's own adopted Korean daughter, in the United States. We learn that as an ineligible older bachelor Hata has bribed a social services agency in order to adopt Sunny, with whom he has a deeply troubled relationship. Their estrangement—at one point, he forces Sunny to have an abortion against her will—not only repeats the unresolved violence and responsibility he harbors toward the murdered K but also recapitulates his own disavowed identity as a Korean adoptee.

As Lee displaces Hata's meticulously normalized performance as exemplary Japanese immigrant, doctor, adoptive father, and Asian American citizen in the United States to disclose his checkered history as Korean adoptee cum Japanese imperial soldier, possible war criminal, and traumatized witness in Asia, an interconnected history of trauma and violence emerges to refuse any radical break of the present from the past. Although Hata tries vigilantly to remain inured to his past, he is slowly forced to reassemble the bits and pieces of his history, like those of K's severed body parts he gathers to bury, to give an account of himself, and to confront his resplendent bad faith. History, however, is not on his side.

The remarkable transformation of Hata from former Japanese wartime enemy into Asian American model minority citizen demands the performance of a specific kind of racialized identity—in gratitude for the "gift of freedom," to return to Mimi Thi Nguyen's loaded phrase—that reinforces the benevolence of the US nation-state as an altruistic, reforming force in international politics and liberation.[85] Hata eagerly obliges. Yet as each of his studiously performed identities unravels, they collectively come to expose the brutal global connections that bind together US and Japanese empire in the face of total war. To return to Fujitani's formulation, as Hata transforms from inhuman Korean colonial into humanized Japanese soldier in Asia, the Japanese enemy similarly transforms into the Asian American soldier and, subsequently, model minority friend in the United States. These connections expose a continuous history of passing, bad faith, and contingent inclusion for exclusion on both sides of the Transpacific. They define Hata's transnational existence even as they reveal a culture of complicity and racial exceptionalism between the two nations, once enemies and now allies, in the aftermath of atomic destruction and occupation.

Hata's "gesture life" in Japan and the United States marks, then, not a break but a continuity between past and present whereby the compliant figure of the Asian American model minority citizen—the anodyne image of immigrant acquiescence and success—comes to displace the violated figure of the inhuman Korean comfort woman who nonetheless remains embedded in Hata's (un)consciousness as a persistent and haunting voice emanating from the forests of Burma. From this perspective, the figure of the Asian American immigrant thus functions as a palimpsest of the disavowed histories of violence across the Transpacific that the victorious nation-state would seek to bury, exerting significant pressure on narratives of benevolence and reform defining myths of American exceptionalism, Cold War freedom, and postwar racial inclusion in an ever-expanding union of "We the People." Commenting on such "traveling memories," Yoneyama writes, "Transnational minorities' memories are never fully in alliance with the dominant national history and memory, yet they are constantly imperiled by nationalizing forces that, through domesticating and assimilating excess knowledge, threaten to produce a seamless narrative of national self-affirmation and innocence."[86]

From a wider political perspective, I argue that the (dis)connections exposed by Hata's traveling memories are mobilized through the language of the human and the civil—through the separation of human right from civil rights—that adhere to dominant national histories on both sides of the Transpacific. That is, just as the figure of the devastated human emerging from the atomic ruins of Hiroshima produced a closed circuit of national victimhood and innocence meant to avoid any reckoning with Japanese war crimes in Asia, the category of the civil performs a similar type of ideological work in the United States.

Throughout his various speeches and writings, Malcolm X lamented the cleaving of the civil from the human in postwar US politics: the separation of civil rights in the United States from human rights on the postwar global stage. For example, in "The Ballot or the Bullet," an address delivered at Cory Methodist Church in Cleveland on April 3, 1964, Malcolm X stressed to his African American audience, "No one from the outside world can speak out in your behalf as long as your struggle is a civil-rights struggle. Civil rights comes within the domestic affairs of this country. All of our African brothers and our Asian brothers and our Latin-American brothers cannot open their mouths and interfere in the domestic affairs of the United States. And as long as it's civil rights, this comes under the jurisdiction of Uncle Sam." Malcolm X's lack of faith in the ballot—that is, in political representation under US democracy—and his abiding insistence on Black self-determination—by force of bullet, if necessary—reflected his deep skepticism toward citizenship as the privileged category guaranteeing racial inclusion under civil rights and a concomitant turn to a global Black nationalism based on common struggles for decolonization in Africa, Asia, and the Americas.

In contrast to Arendt, who posited the category of the human in fascist Europe as politically empty without the legal guarantees of citizenship as "the right to have rights," Malcolm X's trenchant dismissal of civil rights interrupted a long history of American exceptionalism by exposing the category of US citizenship as politically empty without the recognition of the humanity of the Black man, and that of his Asian and Latin American brethren as well. In his *Autobiography*, Malcolm X elaborates:

> Yesterday I spoke in London, and both ways on the plane across the Atlantic I was studying a document about how the United Nations proposes to insure the human rights of the oppressed minorities of the world. The American black man is the world's most shameful case of minority oppression. What makes the black man think of himself as only an internal United States issue is just a catch-phrase, two words, "civil rights." How is the black man going to get "civil rights" before first he wins his *human* rights? If the American black man will start thinking about his *human* rights, and then start thinking of himself as part of one of the world's great peoples, he will see he has a case for the United Nations.[87]

Malcolm X's insistence on situating African American struggles in a larger international order marked a particular cleaving of the civil from the human. If civil rights served to isolate African Americans from other "oppressed minorities of the world," foreclosing the possibility of Third World solidarities and coalitions, they also worked to interpellate African Americans into an ideology of US exceptionalism that was, in fact, a primary source of their subordination.

Notably, while Malcolm X turned to the discourse of human rights as a privileged vehicle for the fulfillment of Black self-determination, the claim to human rights, as Samuel Moyn points out, is an individual as opposed to collective affair.[88] That is, the postwar rise of international human rights emerged with the aim to empower individuals against the violence of the sovereign nation-state. This project, however, was at odds with decolonization movements in Africa, Asia, and Latin America, which sought precisely to build robust and sovereign postcolonial societies. From this perspective, decolonization is not a human rights movement, and human rights cannot easily function as a political panacea to the limitations of the civil. However, if one defines human rights as the failure of European sovereign states to protect individual rights of their colonial subjects, then decolonization becomes in fact an enduring symptom of the problem of reparations, the human, and human rights.

Whether one puts the human before the civil and citizenship, as in the case of Malcolm X's United States, or the citizen before the human, as in the case of Arendt's fascist Europe, it is important to recognize that these two terms have historically functioned in tandem with one another through a long history of liberalism as well as racial capitalism, exploitation, and domination. On the one hand, the language of the civil and citizenship has served to interpellate racial minorities into ideologies of American exceptionalism, while obscuring how citizenship has served as a privileged legal category by which Indigenous groups and other people of color in the United States have been systematically divested of their "right to have rights." On the other hand, the language of the human and human rights has served to interpellate colonized populations into universalist ideologies, while obscuring how the concept of the human has served as the privileged term by which the peoples of Africa and Asia have been consigned to the waiting room of history as uncivilized subjects of development. In this context, Malcolm X's observations about the civil and the human implicitly connect these two critiques, resituating conventional origin stories of human rights as a response to European fascism and the Holocaust by embedding them in a much longer history of colonial expansion, conquest, and dispossession binding Africa, Asia, and the Americas.

Moreover, Malcolm X's arguments encourage us to consider how the language of civil rights seeks to indemnify the US nation-state from charges of human rights abuses both domestically and internationally. Domestically, an exceptionalist narrative of the United States as founded on the rule of law rather than on the rule of kings reinforces an understanding that any national transgression of rights in the United States must be conveyed in the language of civil rights, not human rights, violations. Similarly, on the international level, an ascendant postwar US nation-state can only be putatively the benevolent enforcer, not violator, of human rights and the neutral adjudicator of human rights abuses *elsewhere*.

Aligning this cleaving of the civil and the human more specifically to our discussion of the Transpacific, historians such as William Borstelmann, Dudziak, and Odd Arne Westad point out how struggles between capitalist (US) and socialist (USSR and China) visions of postwar freedom on the global stage directly influenced the development of "Cold War civil rights" for racial progress in Jim Crow America.[89] Under an international lens that magnified the contradictions between US espousals of anticommunist freedom abroad in the face of institutionally sanctioned racial segregation at home, US lawmakers were forced to act as its leaders sought to sell democracy and capitalist freedom to the decolonizing world. Other scholars, including literary scholar A. Yumi Lee and historian Monica Kim, note how civil rights and inclusion for racial minorities in the United States were predicated on soldiers of color enlisting in projects of US empire and militarism in Cold War Asia.[90] (Indeed, the US military was first desegregated during the Korean War, well before the 1954 *Brown v. Board of Education* and 1967 *Loving v. Virginia* Supreme Court rulings outlawed legalized segregation in US civil society.) This instrumentalizing of race for empire, of course, finds an earlier historical precedent in the "voluntary" induction of interned, age-eligible Japanese American men into the US Army during World War II.

It is important to remember that numerous first-generation Japanese American immigrants—barred from naturalization and citizenship by US exclusion laws—were dispossessed of their property and rights, not unlike the displaced and stateless refugees of Arendt's Europe. Nonetheless, postwar Japanese American redress movements for internment were insistently pursued in the name of civil rights. Ironically, as the historian Mae Ngai argues, activist groups such as the Japanese American Citizens League agitated for apology and reparations by deploying the rhetoric of undivided loyalty that flew in the face of transnational affiliations and bicultural connections defining the immigrant experience in America.[91]

President Ronald Reagan signed House Resolution 422, also known as the Civil Liberties Act of 1988. It provided reparations of $20,000 to each former internee still living while offering a national apology for the "grave injustice done to both citizens and permanent residents of Japanese ancestry by the evacuation, relocation, and internment."[92] One of the few successful reparation movements in US history, redress for internment reinforced a narrative of civil rights, liberal inclusion, racial freedom, and a self-correcting democratic nation-state.

On the domestic level, the transformation of Japanese enemies into Asian American model minorities functioned to discipline African Americans, whose lack of social and economic mobility was characterized as a personal failure rather than the result of ongoing structural racism and discrimination. On the international level, this transformation worked to erase a long history of inhuman violation connected to both Japanese and US empires in the Transpacific—a history whose ethical dimensions are richly explored by Ishiguro and Lee. Indeed, the domestic configuration of racial progress under the rubric of civil rights insistently delinked the problem of racial justice at home from violations of human rights in the space of Cold War Asia. To return to Klein, the language of the civil works here to exceptionalize Japanese Americans as good racial subjects, model minority citizens worthy of repair, in contrast to African Americans and comfort women, who are produced as bad racial subjects, inimical to projects of national rehabilitation and uplift and thus outside circuits of consideration and care. In Japan, no such redress for the catastrophic violence that the United States perpetrated against Japanese civilians was thinkable. The sovereign integrity of the victorious US nation-state, as I argued above, vitiated any acknowledgment of response or responsibility.

In *Haunting the Korean Diaspora*, sociologist Grace Cho marks September 1945 as the moment when Japanese colonial rule of South Korea passed into American jurisdiction and hence "signaled the transition between the system of sexual slavery set up for the Japanese Imperial Army (the comfort stations) and the system of camptown prostitution set up for the U.S. military (*gijichon*)."[93] In this way, the figure of the inhuman comfort woman as well as unresolved colonial legacies of sexual slavery are passed down from one empire to another. Just as Lee configures love in *A Gesture Life* not as the resolution to but as the ground for violence and repair, we must similarly evaluate the dialectic of the human and the civil not as the ground for but as the persistent and unresolved problem of a politics of redress haunted by the erasure of the inhuman and a failure of ethical response and responsibility.

I end this chapter by turning briefly to Lee's third novel, *The Surrendered* (2010), which extends the author's exploration of violence in the historical context of the Korean War and through the eyes of three interconnected protagonists: June Han, a Korean war bride who now lives alone in New York City, having immigrated to the United States following the death of her entire family and unsuccessful placement into a Korean orphanage; Hector Brennan, a former American GI and now down-and-out janitor in Fort Lee, New Jersey, who fought in the Korean War and is the father to June's estranged son Nicholas; and Sylvie Tanner, the wife of an American Christian missionary who runs the orphanage where June was placed and Hector worked and whose own missionary parents were killed by Japanese soldiers during their occupation of Manchuria.

The Surrendered not only traces an extended history of conflict across twentieth-century Asia but also connects this carnage to prior violence in Europe, Africa, and the Americas. As a Korean orphan and a subsequent war bride, June embodies two paradigmatic figures of humanitarian rescue in the context of total war. June indexes a history of repair that the novel probes through its numerous scenes of lurid violence and a recurrent leitmotif, a book within a book that circulates across the time and space of Lee's epic novel: Jean-Henri Dunant's memoir, *A Memory of Solferino*, published in 1863.[94] After several decades of no contact, and rapidly declining from her spreading stomach cancer, June finds Hector and enlists him to aid in her search for their lost son, who has disappeared in Europe. The novel ends in Solferino, Italy, the site of a historic battle during the Second Italian War of Independence in which allied French and Italian forces defeated the Austrians, contributing to the eventual unification of a sovereign Italy. Though Lee does not mention it directly, Solferino is also the birthplace of modern humanitarianism.

Dunant's *A Memory of Solferino* recounts the historic battle that took place on June 24, 1859. In *The Surrendered*, a tattered copy of *A Memory of Solferino* is passed down from Sylvie's missionary parents to their steadfast daughter in China, stolen by the teenager June when she burns down the Korean orphanage (resulting in Sylvie's accidental death), and finally taken by Nicholas when he disappears from New York City and his mother's life. Dunant (1828–1910), a Swiss businessman and representative for a corn-growing and trading company in colonial Algeria, had traveled to Italy for an audience with the French emperor Napoleon III, who was stationed with his army at Solferino. Dunant hoped to secure land and water rights for his company from French colonial authorities through a direct appeal to the emperor.

Dunant thus happened upon the epic battle, one mobilizing three hundred thousand soldiers. In the aftermath of the fighting, he was stunned by the carnage and desolation: forty thousand dead, wounded, and dying remained scattered on the battlefield. Under the motto *Tutti fratelli* (All are brothers), Dunant helped to organize local civilians to provide assistance to the injured combatants regardless of the side on which they were fighting. Upon his return to Geneva, Dunant began to advocate for the establishment of an international organization to deliver neutral aid in the face of war and conflict. In 1863, he founded the International Committee for the Red Cross (ICRC), and a year later he helped to convene the first Geneva Convention, which laid out universal protocols for care of wounded soldiers in battle. For his efforts, Dunant was awarded the Nobel Peace Prize in 1901.

As *The Surrendered* starkly illustrates through its depictions of graphic violence, Dunant's legacy of repair—of humanitarian aid and relief—has scarcely made a difference. In the aftermath of atomic destruction, projects of total war have come to encompass more and more civilian deaths, alongside those of combatants—a stark reality underscored by Hiroshima and Nagasaki as well as the grim fact that Korea alone lost an estimated 10 percent of its civilian population during the war. Serving as the weighty, if distant, referent for Lee's main story about the Korean partition and its proliferating chain of traumatic aftereffects across three characters and continents, Solferino raises a comparative quandary concerning universal principles of humanitarian aid in the service of protecting human lives in Europe and elsewhere during conflict that Dunant's book putatively enshrines.

Comparing Dunant's memories of Solferino with those of his own in Korea, as well as the countless wars that have occurred in the interim century, Hector interrogates Sophie about her humanitarian ideals on the eve of her departure and the closing of the orphanage: "Did you think you could come and go so easily? Is this what happens in that precious book of yours? I want to know. I thought it was about showing mercy to the helpless, to the innocent. But I think that book of yours is worthless. In fact, it's worse than that. It's a lie. It's changed nothing and never will."[95] Lee's third novel thus stands as an extended disquisition on the failure of modern regimes of humanitarian aid inaugurated by Dunant's memoir in the context of Asia and the Korean War. "By making Solferino the ending place both for June and for the reader of the novel," A. Yumi Lee observes in her analysis of the novel, "*The Surrendered* enacts a return of the violence of the Korean War to the grounds of Solferino, at once paying homage to and indicting the mission which it birthed."[96]

Contemporary critics of humanitarian aid and rescue point out that the role of the officially neutral, apolitical relief agency serves to alleviate antagonists of many of the burdens of waging war. For instance, as these agencies take on the administrative and financial costs of attending to the sick and wounded without regard to politics, they often heal combatants only to see them return to battle and exacerbate the war's ongoing atrocities. As *New Yorker* journalist Philip Gourevitch writes in his critique of humanitarianism, "At its worst—as the Red Cross demonstrated during the Second World War, when the organization offered its services at Nazi death camps, while maintaining absolute confidentiality about the atrocities it was privy to—impartiality in the face of atrocity can be indistinguishable from complicity."[97] As it turns out, neutral humanitarian relief agencies cannot absent themselves from the politics of history. We return again to the problem of victims and perpetrators and to the predicament of repair as the ground for, rather than the solution to, violence.

In his study on human rights, international relations scholar Stephen Hopgood observes that the ICRC was "the first international human rights organization. It was the secular church of the international. The laws it wrote and the humanitarian activism it undertook were grounded by a culture of transcendent moral sentiment with strong Christian components. At the heart of this was the suffering innocent, a secular version of Christ. In other words, bourgeois Europeans responded to the erosion of religious authority by creating authority of their own from the cultural resources that lay scattered around them. And then they globalized it via the infrastructure that the imperial civilizing project bequeathed to them."[98] In the American Century, it was the United States that lay claim to the legacy of this imperial civilizing project through the "gift of freedom"—through preemptive violence and rollback—that came to define the abject ruins of Korea, Vietnam, and numerous other areas of Cold War conflict. This legacy, as I have argued throughout this chapter, is one circumscribed by the problem of racial reparations and the human that cannot account for the erasure of the suffering of those rendered inhuman by its Euro-American dialectic.

Like trauma, human rights, as Moyn emphasizes, is structured by repetition. It has arisen two times. Appearing in the wake of the Holocaust and the Nuremberg Trials and then disappearing in international politics until its reemergence in the 1970s, the rapid ascension of discourses of human rights as we understand them today, during this second period, is predicated on aligning it with capitalist visions of modernity—that is, predicated on associating its abuses with authoritarian socialist and left-wing decolonizing regimes rather than right-wing, totalitarian fascist regimes such as the Third Reich. As political theorist William Pietz describes it, the "necessary conscience-smoothing exorcism was achieved by affirming the equation of Nazi Germany and Soviet Russia, combined with an historical interpretation of the essential Orientalness of the Russian mentality. The basic argument is that 'totalitarianism' is nothing other than traditional Oriental despotism plus modern police technology."[99] In this manner, just as the civil is hived off from the human, human rights are hived off from the dream of socialist modernity and Third World decolonization—the dream of justice under the long shadows of racial capitalism.

The return of human rights thus created a link between prior colonial and present Cold War discourses that I have traced in chapter 1 from Locke to Klein, and chapter 2 from the Holocaust to Hiroshima, and from the Transatlantic to the Transpacific. In contrast to the dream of human rights as a renewed Enlightenment—"the last utopia," in Moyn's phrase—the enigma of the fictionalized voices that call out for the traumatized Hata, Hector, Sylvie, and June to see what they have done—to realize what they have witnessed—opens up a space in history for a different politics of repair to emerge outside paradigms of national sovereignty and its privileged citizen-subjects. Critical theorist Pheng Cheah argues that Freud's conception of trauma implicitly posits a politics of sovereignty in the form of a bounded subject and a bounded nation prior to their traumatic shattering.[100] The enduring problem of racial reparation and the human in the face of unrelenting war and violence cannot be predicated on struggles for sovereignty—either the sovereignty of the liberal individual, the postcolonial citizen-subject, the European nation-state in Dunant's time, or the postcolonial nation-state in ours. Instead, we must think the politics of repair beyond trauma and outside of sovereignty and the universal human subject altogether. We must focus on the constitutive exposure of the self-same subject and an experimental ethics of recognition and responsibility intended toward the humanity of dispossessed and disposable others. I turn to this experimental ethics in chapter 3, the final chapter of this book.

3

Beyond Sovereignty

Absolute Apology, Absolute Forgiveness

What I dream of, what I try to think as the "purity" of forgiveness worthy of its name, would be a forgiveness without power: *unconditional but without sovereignty*. The most difficult task, at once necessary and apparently impossible, would be to dissociate *unconditionality* and *sovereignty*. —JACQUES DERRIDA, *On Cosmopolitanism and Forgiveness*

When they discovered the ore at Port Radium, which was then sent down south, and they made a bomb from it and destroyed all those many lives in Japan, we weren't even aware about that, even though it came from our lands. We Dene people are a good people, and we don't want them to think it was our fault. I'm sorry for them, and I want to send my respects. I hope that blame won't be put on us because we had no knowledge of what happened in the war. —BELLA MODESTE, Deline uranium widow, in Blow, *A Village of Widows*

To be undone by another is a primary necessity, an anguish to be sure, but also a chance—to be addressed, claimed, bound to what is not me, but also to be moved, to be prompted to act, to address myself elsewhere, and so to vacate the self-sufficient "I" as a kind of possession. If we speak and try to give an account from this place, we will not be irresponsible, or, if we are, we will surely be forgiven. —JUDITH BUTLER, *Giving an Account of Oneself*

We need a new concept of reparations and the human beyond sovereignty—beyond the unconditional sovereignty of the nation-state and beyond the self-sufficient "I" as property and possession. We need a new concept, that is, accountable not only to the fraught political history of colonization, decolonization, the human, and human rights but also to the fraught biopolitical history of the human being—indeed, to the precarity of all creatures and things inhabiting the earth, and to the earth itself. If we are in relation to and undone by one another, if injury is inevitable, and if bad objects are saturated with the colonizing projections of others, then how is such damage ever to be addressed and repaired? In the final analysis, injury is not reparable; only relationships are. And we repair only those relationships that are valuable to us.

Chapter 3 maps a response to the social and psychic quandary of repairing bad objects. I end where I first began in chapter 1 with my analysis of John Locke and Melanie Klein: Indigenous dispossession in the Americas. This chapter investigates the history of uranium mining and, in particular, Little Boy, the atomic bomb detonated by the US military over Hiroshima. Most of the world's uranium supply is mined from Indigenous lands.[1] The uranium sourced for the creation of Little Boy was no exception. It was extracted in part from the territories of the Dene, a First Nations peoples inhabiting the Sahtu region of the Northwest Territories in Canada. On May 16, 1930, prospector Gilbert LaBine discovered silver as well as pitchblende, a combination of radium and uranium ore, on the shores of Great Bear Lake. From 1934 until 1960, when mining operations were shuttered, the Sahtu Dene men worked as "coolies" (a term used to refer to them in the industry) for the Eldorado Mining and Refining Limited Company, which was secretly nationalized as a Crown Corporation in 1942 by the Canadian government as part of their Allied war efforts.[2]

During these years, the Dene men helped to transport radioactive ore from a mine in Port Radium at Echo Bay, westward across the Great Bear Lake to Deline, and then down the Great Bear River farther west for refinement at Fort Norman (map 3.1). From there the ore was sent to Port Hope, Ontario, for further processing and then shipped to the US government for exclusive use in the Manhattan Project. Following the destruction of Hiroshima and Nagasaki, ore from the mine was deployed for the buildup of the US nuclear arsenal during the Cold War arms race, and uranium became Canada's most profitable mineral export. This battle for atomic supremacy and mutual assured destruction is often described as the greatest industrial undertaking in the history of mankind. From numerous perspectives, as I elaborate below, it was certainly the most costly.

Many of the Sahtu Dene men who labored on behalf of this atomic initiative died of cancer.[3] Their families and descendants all suffer from extreme levels of radiation poisoning and elevated rates of malignancy. The land on which the Sahtu Dene live and its ecosystem, mined without their consent, are poisoned by radioactive waste for tens, if not hundreds, of thousands of years. Ignorant

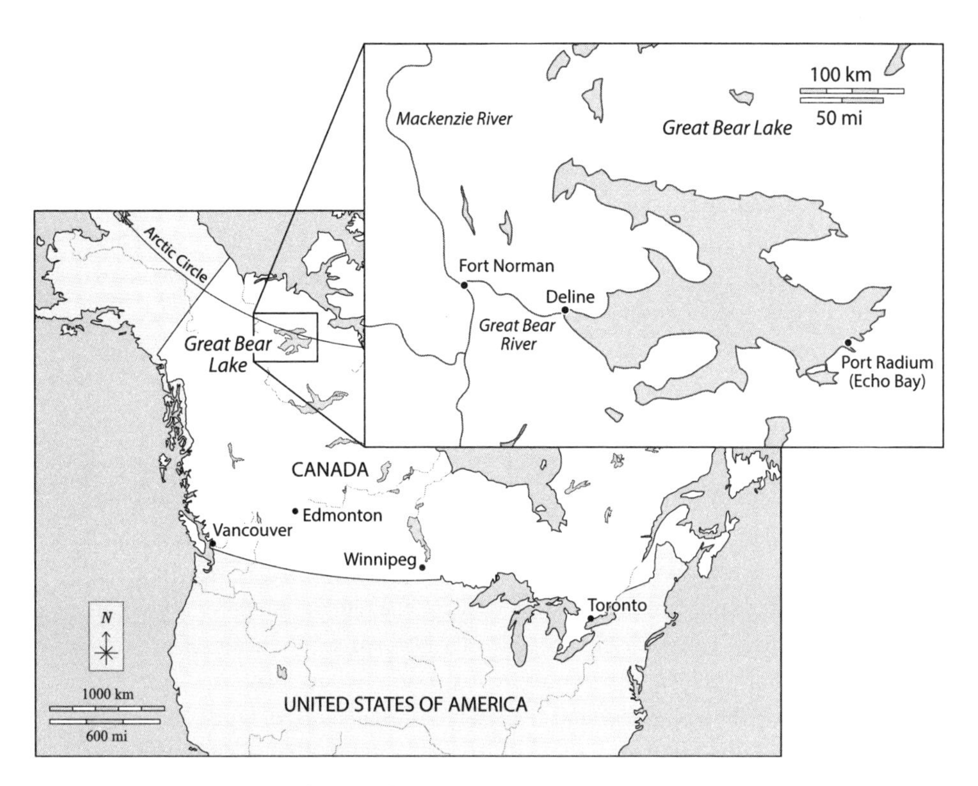

MAP 3.1. Great Bear Lake, Canada

at the time of how their efforts would be applied, and lacking any knowledge of the final destination of the ore, the Sahtu Dene nonetheless felt implicated once they learned of their connection to Hiroshima's fate. In response, they sent a delegation to Hiroshima to apologize.

This chapter considers the implications of this startling act of apology. Few people know that Canada provided much of the uranium for the atomic bombs that were dropped on Hiroshima and Nagasaki. Even fewer are aware of the devastating health effects and environmental toll that uranium mining has exacted on the Great Bear Lake region, the Sahtu Dene, and numerous other Indigenous groups from whose territories radioactive minerals have been extracted. Below I focus on the manifold implications of the Dene's pilgrimage to Hiroshima by extending Jacques Derrida's notion of "absolute forgiveness" to develop a corollary concept: "absolute apology."

If the postwar emergence of international human rights sought, however unsuccessfully and unevenly, to curtail the unconditional sovereignty of the nation-state, this chapter proposes another approach to reparations and the human outside structures of state governance, the drive to install the dominance of man over the earth, and attendant political calculations meant to evade response and responsibility to those who are harmed. It considers what is at stake when the colonized and dispossessed, rather than the colonizer and settler, take the initiative to apologize. The Dene's apology offers a different mode of address to the other. It provides a different model of redress, of response and responsibility—one that seeks to repair those bad objects created yet abandoned by recursive histories of colonial trauma and violence.

To put it otherwise, conventional histories of reparation in colonial modernity retroactively create the figure of the violated human being—that is, a victim and good (colonial) object worthy of recognition and repair. In contrast, the Sahtu Dene's apology interrupts the constitution of this privileged figure. It works to acknowledge unacknowledged others—bad (Indigenous) objects injured by the sovereign state yet undeserving of repair. Indeed, it reworks notions of sovereignty and citizenship as the privileged site of response and responsibility in liberal politics and, in turn, challenges the figure of the human being itself across various geopolitical times, spaces, and scales. The Sahtu Dene's apology undoes the intransigent dialectic that binds "apology" and "forgiveness" with "perpetrator" and "victim." It offers a model of repair that seeks to address unaddressed atrocities perpetuated though disavowed by the sovereign state and its privileged agents of destruction. "You can't really be sorry for something that you don't want to remember, can you?" asks Round Rose in playwright Marie Clement's *Burning Vision*, a 2003 stage drama based on the history of uranium mining among the Sahtu Dene.[4]

Chapter 3 narrates an alternative account of the atomic bombings of Japan and the postwar politics of reparations and the human after nuclear holocaust by connecting the much longer history of Indigenous dispossession in the New World with more recent militarism and violence in Cold War Asia. By thinking together the figures of the Indian and the Asian across the Transpacific, I keep open a space for new constellations of the injured to emerge and for new atrocities to be apprehended—an atomic flash of memory, as Max Horkheimer and Theodor Adorno put it in *The Dialectic of Enlightenment*, on a "fully enlightened earth [that] *radiates* disaster triumphant."[5]

This chapter proceeds in three sections. It begins with a history of uranium mining as well as the uneven social and ecological effects of nuclear fallout, what anthropologist Joseph Masco describes as the least recognized aspect of the Atomic Age: toxic colonialism.[6] The concept delineates the disproportionate and deleterious effects of uranium mining on communities and creatures, humans and nonhumans alike, who inhabit Indigenous lands across the planet and from whose territories radioactive minerals are extracted. Uranium's devastating effects, often immediate, are also extended over vast periods of time, indexing what postcolonial scholar Rob Nixon labels "slow violence," the gradual and often hidden harms of toxicity on disenfranchised and abandoned communities.[7] Second, this chapter offers an analysis of the Sahtu Dene's act of apology by examining Peter Blow's 1999 documentary *A Village of Widows*, which recounts the Dene's gesture of contrition in Hiroshima. Finally, the chapter concludes with a critique of normative state politics of apology and forgiveness: the intransigent division of victim from perpetrator, the persistent break between truth and reconciliation, and the putative separation of human from nonhuman that undergirds conventional accounts of postwar human rights and the politics of reparation. These assumptions buttress an Enlightenment faith in sovereignty, scientific advancement, and human progress, one interrupted by the Sahtu Dene's act of absolute apology—"unconditional but without sovereignty," as Derrida suggests in the epigraph above—and one that opens onto an alternative political as well as psychic terrain for reparations and the human.[8]

Toxic Colonialism

The Cold War is largely recounted as a struggle between two competing empires, with the United States and the USSR jockeying for capitalist and socialist hegemony, respectively, in the postwar period. In turn, conflicts between these two superpowers across the decolonizing world undermined the sovereignty and self-determination of many Third World nations. This was especially so in Cold War Asia. Throughout the region, a series of hot wars and partitions created a stream of stateless and displaced peoples. War, militarism, occupation, covert operations, and containment shattered dreams of decolonization, sovereignty, and nonalignment that challenged competing capitalist and socialist visions of national development and progress. At the same time, unending conflict also forestalled any reckoning with overlapping histories of colonialism and violence in East and Southeast Asia. These histories remain unresolved to this day.

The bombings of Hiroshima and Nagasaki inaugurated the Atomic Age. In retrospect, they also marked a decisive moment in the Anthropocene. Coined in the 1980s by Eugene Stoermer, an American ecologist, and Paul Crutzen, a Nobel Prize–winning Dutch chemist, the term *Anthropocene* describes an era in which collective human activity came to exert a geophysical force on the planet. Representing our most recent attempt to tell the story of humankind's disastrous impact on the planet as one of unwitting destruction and deferred responsibility, the Anthropocene is typically indexed to the invention of James Watt's steam engine in 1776, the ever-increasing burning of fossil fuels, and the concomitant emissions of carbon that have altered the earth's atmosphere. Relatedly, the great acceleration of greenhouse gases and rising temperatures (often described as the "hockey stick effect") is traced to 1945 and the intensification of postindustrial globalization in the West. This great acceleration must also be considered in relation to nuclear fallout.

The term *fallout*, which appeared in the English language soon after the nuclear detonations over Hiroshima and Nagasaki in 1945, refers to the radioactive toxic debris that blanketed and continues to blanket the entire planet's atmosphere in the wake of the atomic bombings as well as subsequent nuclear arms production, use, and testing across the earth.[9] Fallout might also serve as a metaphor for the escalating emergencies represented by the Anthropocene and climate change—an atmospheric event with far-reaching and unknown consequences. Masco argues that, together with Cold War political concerns for national sovereignty and security, the term has shifted us from a "global" to a "planetary" imaginary oriented toward biopolitical collapse. Although depictions of a spherical globe are long-standing, he writes, the "specific attributes of being able to see the entire planet as a single unit or system is a Cold War creation," indexing planetary imaginaries of finance, technology, and international relations, along with geology, atmosphere, biosphere, glaciers, and oceans as one singular totality.[10] Paradoxically, we now live in an age of fallout that not only relies on the national security state for the finances, technologies, and relations that form the conditions of possibility for seeing the planet in this particular manner but also decidedly exceeds the nation-state as the proper locus of adequate redress.

From another perspective, while the production and testing of nuclear weapons were (and continue to be) insistently aligned with the sovereign priorities of national security, both their material conditions of production and their long-term social, ecological, and planetary effects on the environment lie beyond the framework of any sovereign nation-state or its privileged citizen-subjects. "We do not have political systems," Masco notes, "operating on the right scale to address truly planetary problems."[11] These sovereign and planetary dilemmas must be traced not only to Hiroshima and Nagasaki but also to earlier histories of dispossession and extractive capitalism from Indigenous lands: histories that form the conditions of possibility for the emergence of Cold War conflict, the nuclear arms race, and its manifold effects. As the irreversible consequences of fallout exceed the human imagination and hence our capacity to respond politically, they pose unprecedented challenges to the problem of repair, human and otherwise, giving rise to the prospect of irreversible damage and irreparable injury.

To begin, the entire production cycle of nuclear weapons cuts across national, transnational, global, and planetary scales, decisively turning our attention in stark ways to the biopolitical implications of the reordering of human and nonhuman life. The uranium sourced for the Manhattan Project came from three sites: the Belgian Congo, northwestern Canada, and the southwestern United States. As with the Soviet Union, which relied on its internal colonies of Kazakhstan and Uzbekistan for its national uranium supply, the United States began intensive exploration for domestic deposits of uranium immediately after the conclusion of World War II. Because the US government, the largest uranium producer in the world during this period, hoped to achieve nuclear independence under the shadows of Cold War conflict, the ore it consumed came increasingly from mines in the Colorado Plateau. (Significant uranium deposits were also extracted from mines in Wyoming and South Dakota.) Known as "Four Corners," the region where the states of Arizona, Colorado, New Mexico, and Utah come together, the Colorado Plateau consists of territories belonging to the Hopi, Navajo, Pueblo, and Ute peoples.

Since the original discovery of uranium in Arizona in 1941, for instance, more than a thousand uranium mines opened between 1944 and 1989 on leased territories of the Navajo Nation. According to environmental scholar Valerie Kuletz, during this period nearly 4 million tons of ore were extracted from Navajo sites in Arizona, New Mexico, and Utah.[12] In an even shorter span of time, from 1952 to 1981, an astonishing 22 million tons of uranium ore were mined from Laguna Pueblo lands near Paguate, New Mexico, which contain some of the richest supplies of uranium in the region as well as the largest open-pit uranium mine in the world (the Jackpile-Paguate Mine). At the same time, despite national dreams of uranium autonomy, the US military continued to rely on foreign uranium deposits, including those from Sahtu Dene and other First Nations territories in Canada.

The workforce for this colossal scientific endeavor has been similarly transnational. For instance, the Manhattan Project—headquartered at Los Alamos National Laboratories (LANL) in New Mexico—involved a network of laboratories in the United States, Canada, and the United Kingdom that totaled 130,000 employees. LANL employed low-level miners and laborers of Indigenous and Mexican descent as well as high-level foreign scientists of divergent backgrounds and nationalities.[13] Notably, even before Germany's official surrender in World War II, prominent physicists, along with nuclear weapons technologies, were surreptitiously ferried out of Germany to be redeployed in competing Allied efforts for the continuing development of postwar atomic initiatives.[14]

Between 1940 and 1996, the US government spent over $5.8 trillion to construct seventy thousand nuclear warheads, making the US atomic arsenal one of the largest industrial enterprises in human history. According to Ojibwe activist Winona Laduke, the nuclear industry "is perhaps the most highly subsidized industry in the United States[,] . . . dol[ing] out some $97 billion in subsidies for the nuclear industry between 1948 and 1992."[15] During these decades, the United States conducted a total of 1,149 nuclear test detonations—in the continental United States (Alaska, Colorado, Mississippi, Nevada, and New Mexico), the Atlantic Ocean, and in the Pacific Proving Grounds. These transpacific areas include Bikini Atoll, part of the Republic of the Marshall Islands in Pacific Micronesia that became a US protectorate following the surrender of Japan. (Earlier it had been a German protectorate.) US nuclear facilities spread out over a total landmass of over thirty-six thousand square miles, larger than the combined states of Massachusetts, New Hampshire, Vermont, Maryland, and the District of Columbia.[16]

In similar fashion, the Soviet Union and subsequently the United Kingdom, France, China, India, Pakistan, and, most recently, North Korea have each tested their own atomic devices. Indeed, almost all nuclear-armed nation-states assert their sovereign and superpower status through the very public display of weapons tests.[17] As the international legal theorist Antony T. Anghie observes, we "live in a world governed by the empire of nuclear weapons and the rules it prescribes. The edicts of this empire are translated and elaborated for us into the commanding and incontestable rules of security, self-defense, and sovereignty by experts in strategic studies and international relations."[18]

With the five major nuclear powers today (the United States, Russia, the United Kingdom, France, and China) all occupying permanent seats on the United Nations Security Council, this nuclear empire of mutual assured destruction is entrenched into the very structure of postwar international law. Neither the Treaty on the Non-proliferation of Nuclear Weapons, which was inaugurated in 1970 and requires all nuclear member states to engage in good-faith negotiations toward disarmament, nor the conclusion of Cold War hostilities in 1989 with the fall of the Berlin Wall and the dissolution of the Soviet Union has deterred the production and development of these weapons of mass destruction. In fact, after a mid-1980s slump in production and profits, nuclear arms manufacturing was renewed with particular vigor after the 9/11 attacks on New York City's World Trade Center and the advent of the now-interminable "war on terror." Since 2005, a new uranium boom has emerged, with a concomitant spike in ore prices and the reopening of previously closed mines across the world, including those on Dene territories.[19]

Anghie points out that in confronting North Korean atomic testing today, the global community faces the "bizarre but logical" outcome of its own founding principles: that "sovereignty *is* nuclear weapons"—a principle that "dispenses with even a minimal concern for the everyday welfare of its [the global community's] people."[20] If, in the aftermath of Hiroshima and Nagasaki, sovereignty has come to be defined through nuclear weapons, the fallout from nuclear weapons production and testing across the world nonetheless exceeds any dream of unconditional national inviolability. Framed as a resolutely political question of national sovereignty and security, yet inescapably transnational in their material conditions of production, use, testing, and effects, nuclear arms and their biopolitical consequences are undeniably a planetary problem of vertiginous entanglements implicating human and nonhuman ecologies alike.

More specifically, the planetary effects of fallout demand analysis on multiple levels of the ecosystem, the organism, and even the cell, reconfiguring traditional conception of human and nonhuman relations on multiscalar levels.[21] Today there are no beings—flora or fauna, lakes or oceans—on planet Earth that do not exhibit heightened levels of radiation exposure due to fallout. Like climate change, fallout disregards geopolitical boundaries and borders. The mining, milling, refining, and enrichment of radioactive minerals generate lethal waste products of unimaginable and unpredictable toxicity, with no commensurate human technologies to safely contain them or dispose of them.

Uranium-235, for instance, which was used in the construction of Little Boy, has a half-life of 704 million years—a paltry figure in comparison to that of uranium-238, used in the production of Fat Man (dropped on Nagasaki), which has a half-life of 4.46 billion years. The products generated from the enriching of uranium, including plutonium and neptunium, are heavier elements than uranium and far more volatile. Neptunium's half-life is approximately 2 million years, while plutonium's half-life is a scant 24,000 years. Both neptunium's and plutonium's particles are lethal if inhaled in even microscopic amounts. Even when sealed and buried, they can seep into groundwater, leak into underwater crevices, and form high-pressure gases that emit corrosive energies onto the earth's surface.[22] The task of nuclear waste disposal requires a financial expenditure and scientific commitment exceeding that of the original Manhattan Project—and we are only some seventy-five years into this nuclear experiment.

Such figures and comparisons defy the human imagination. Indeed, the half-lives of some radioactive materials exceed the cumulative time that the human species has inhabited planet Earth, radically reconfiguring the temporalities of trauma and injury. Notably, the time of fallout is retrospective and anticipatory, marked by a lag between the environmental event and the recognition of its long-lasting consequences. Even more, the strange temporality of fallout marks a significant psychic shift that reconfigures political practice in our Atomic Age. "In the name of commerce or security," Masco observes, "consequences are loaded into an uncertain future and thus expelled from the realm of formal political discourse." Fallout is "therefore understood primarily retrospectively, but it is lived in the future anterior becoming a form of history made visible in negative outcomes."[23]

With regard to this unimaginable future, art historian Julia Bryan-Wilson has explored the problem of how to create nuclear signage that warns unborn generations, however advanced or primitive, of the unrevealed dangers of radioactive waste sites. A striking example of this temporal conundrum, the dream of a universal sign system—or, more accurately, the impossibility of a universal sign system that will endure across all time and space—underscores the ways in which atomic fallout is simultaneously a crisis that has passed us by and a crisis yet to come. This temporal paradox, this future anterior of disaster, marks a breakdown in communication itself. It is a political, social, scientific, and aesthetic predicament engendered by man, yet a catastrophe in excess of human agency to redress.

Like climate change, the dangers of uranium mining far exceed humans. They reconfigure hierarchical relations between human and nonhuman, organic and inorganic matter, animate and inanimate, subject and object. To be sure, the sovereign will of Enlightenment man—the drive to approach the earth as an object to be mapped, dominated, and owned in its entirety—meets its end in fallout and environmental collapse. Radioactivity underscores that objects have hidden reserves that cannot be known, that are beyond human capacity, comprehension, and wonder.

Consider, for example, a passage from John Hersey's "Hiroshima," analyzed in the previous chapter. Miss Sasaki, returning to the ruined city nearly five weeks after the destruction of Little Boy, encounters a paradoxical scene of botanical terror and health that both "horrified and amazed" her: "Over everything—up through the wreckage of the city, in gutters, along the riverbanks, tangled among tiles and tin roofing, climbing on charred tree trunks—was a blanket of fresh, vivid, lush, optimistic green; the verdancy rose even from the foundations of ruined houses. Weeds already hid the ashes, and wild flowers were in bloom among the city's bones."[24] In similar terms, writing about the verdant proliferation of plant life throughout New Mexico's uranium hot zones, Masco observes that what is utterly surprising and unanticipated about the surrounding vegetation is not that it is highly radioactive but, rather, that it appears to be thriving. This is one paradox of the strange duality of the nuclear age: "that contamination, and the possibility of mutation, can travel hand in hand with visible signs of health and prosperity."[25]

In a series of prescient essays on the specter of climate change, postcolonial historian Dipesh Chakrabarty notes that the ecological crises engendered by the age of the Anthropocene present us with a new universal version of the human as "species being," one acting with a geophysical force on the planet, and one in distinct tension with not only the universal Enlightenment human of liberal rights and representation but also the particular postcolonial human, endowed with "'anthropological difference'—differences of class, sexuality, gender, history, and so on."[26] In privileging this new universal image of the climate-changing human in the age of the Anthropocene (*anthro-* of course meaning "human" in Greek), Chakrabarty's warning nonetheless de-emphasizes two critical aspects of fallout: first, the overlapping relations of human and nonhuman ecologies that radiate from uranium mining; and second, the radically uneven effects of nuclear fallout's immediate as well as slow violence, which are borne disproportionately by Indigenous, colonized, and racialized communities burdened with colonialism's "anthropological difference." As uranium's toxicities are asymmetrically distributed, fallout has already affected particular communities and places in unequal, tangible, and disastrous ways: the Sahtu Dene in northwestern Canada, numerous Native American communities in the US Southwest, and Indigenous populations in Australia and Micronesia.[27]

Ironically, while nuclear arms production has been rationalized as deterring total war, thereby associating the perverse buildup of nuclear arsenals with peace and progress, it has already devastated many Indigenous communities and environments across the planet.[28] In New Mexico, for example, which produced half the national uranium supply during the Cold War, the entire production cycle of nuclear production negatively impacts Native American communities who have inhabited the land for generations: the uranium ore is mined from Native lands; the bombs are designed on Native lands, at Los Alamos National Laboratory and Sandia National Laboratory, and tested on Native lands, at the White Sands Missile Range; the nuclear arsenal is stored on Native lands, at Kirtland Air Force Base near Albuquerque; and the nuclear waste products are redeposited on Native lands, at the Waste Isolation Pilot Project in Carlsbad (map 3.2). Apart from the Hopi and Zuni nations, all sixteen eastern Pueblo territories in this region are located within the immediate ambit of a US nuclear facility. These nuclear structures overlap with Indigenous ge-

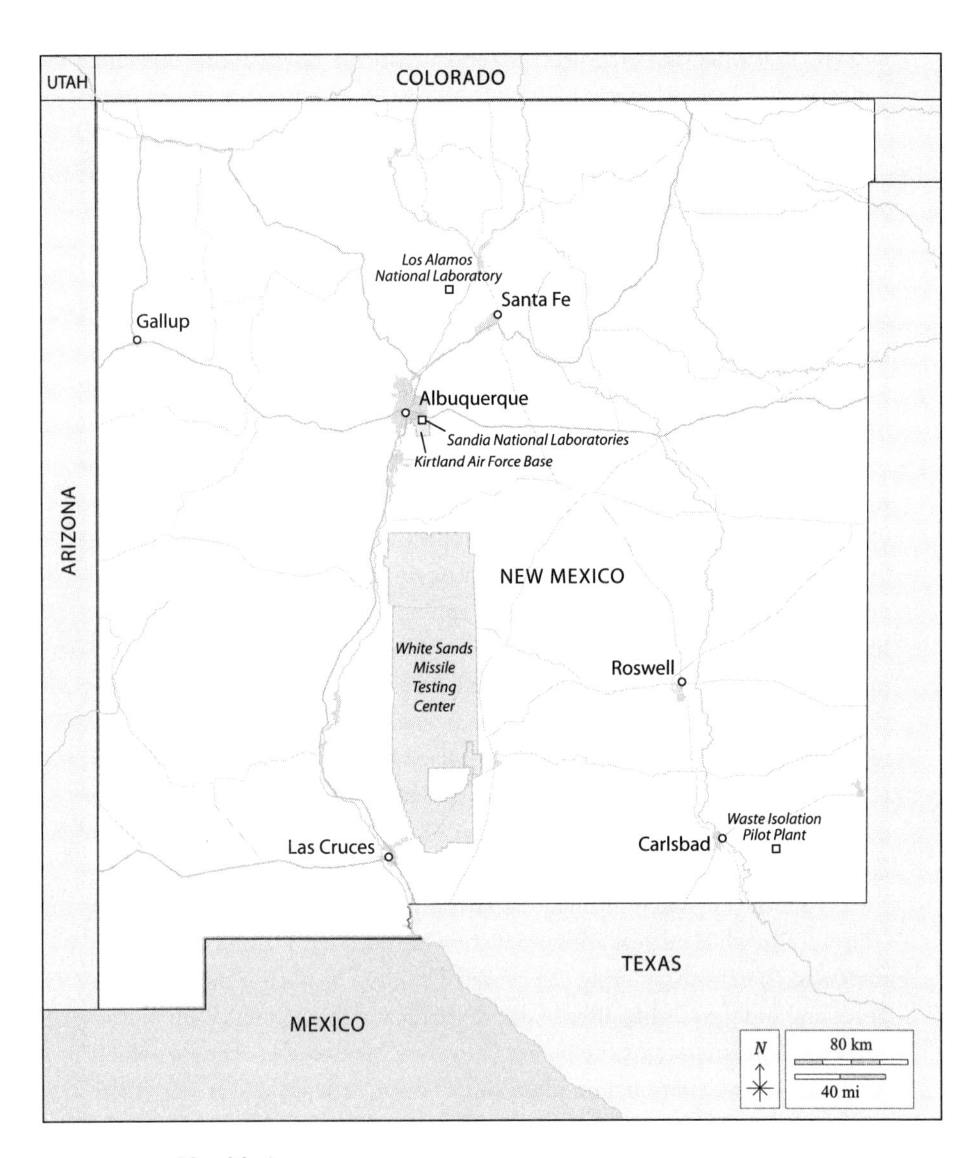

MAP 3.2. New Mexico

ographies in every phase of atomic production, testing, and use. Indeed, when US military planners sought out a peripheral space in 1943 for the construction of the Manhattan Project, "they colonized the geographical center of Pueblo and Nuevomexicano territories, engaging cosmological orders that identify the northern Rio Grande valley as quite literally the center of the universe."[29]

Similar to the Sahtu Dene, Indigenous communities in New Mexico experience significantly higher rates of cancer than other ethnic groups—a major shift from the first decades of the twentieth century, when cancer was such a rare occurrence among Native Americans that some medical specialists believed them to be immune to the disease.[30] Since the dawn of the nuclear age, however, cancer has become a leading cause of mortality among these communities. Environmental studies scholar Traci Brynne Voyles has examined the astronomical rates of cancer deaths among former uranium miners from Navajo Nations. "By the mid-1980s," she writes,

> miners contracted lung cancer at rates 56 times higher than the national average, and had an average life expectancy of only 46 years. Rates for stomach cancer were 82 times the national average. Miners were more than 200 times more likely to get liver cancer, almost 50 times more likely to get prostate cancer, and over 60 percent more likely to have cancers of the bladder or pancreas. Nor were cancers the only health problems among former miners and their families: researchers also found increased incidents of tuberculosis, fibrosis, silicosis, and birth defects, all linked to exposure to uranium from mines and mills.[31]

Radiation-related illnesses are now prevalent in numerous Navajo populations, claiming miners and nonminers alike. Through all these examples, we witness a new atomic vision of Manifest Destiny and westward expansion that apprehends the southwestern United States as a barren landscape devoid of inhabitants—indeed, eviscerated of them.[32]

It was precisely such a logic of land bereft of human inhabitants or significant life (*terra nullius*) that similarly led the British government to repeatedly conduct nuclear testing in Maralinga, a region in the Woomera Prohibited Area of South Australia.[33] More recently, the unsuccessful legal claim of the Marshall Islands in the International Court of Justice against US nuclear testing in Bikini Atoll underscores proliferating connections among uranium mining, toxic colonialism, and longer histories of Indigenous dispossession across the Transpacific. Anghie remarks, "Writing in 1831 of the tragic plight of the Cherokee nation whose members had desperately sought the intervention of the US Supreme Court to uphold the treaties between the United States and the Cherokee whereby the former solemnly promised to protect the latter, Chief Justice Marshall observed, '[i]f courts were permitted to indulge their sympathies, a case better calculated to excite them can scarcely be imagined.' The plight of the people of the Marshall Islands can surely be described in similar terms."[34] The closing lines of Marshall's opinion in *Cherokee Nation v. State of Georgia*, as I discussed earlier, are a prescient and chilling invocation of the future anteriors of fallout nearly a hundred years before uranium's discovery on, extraction from, and testing on Indigenous lands in the United States and across the world: "If it be true that the Cherokee nation have rights, this is not the tribunal in which those rights are to be asserted. If it be true that wrongs have been inflicted, and that still greater are to be apprehended, this is not the tribunal which can redress the past or prevent the future."[35]

Marshall's closing lines are an uncanny reminder of the temporal paradoxes of uranium—of wrongs inflicted in the past and still greater harms to be apprehended in an unknowable future. This is the time and space of Indigenous dispossession, which is to say the time and space of colonial modernity. In the face of fallout and toxic colonialism, this has become our time as well. Places like the Marshall Islands, Maralinga, New Mexico, Great Bear Lake, Hiroshima, and Nagasaki exemplify the problem of racial reparations and the human in the face of the long histories of colonial settlement, dispossession, and enclosure. As the proliferating crises of nuclear entanglements and ecological disaster, the wayward effects of extractive capitalism and accumulation, reconfigure histories of human injury in unfathomable ways, we are left with legacies of state sovereignty and citizenship wholly inadequate to our current state of emergency.

The coming together of political and biopolitical devastation in the postwar age of fallout and climate change, more generally, gives rise to what environmental studies scholars Ashlee Cunsolo and Neville R. Ellis have labeled "ecological grief": the anticipated but unknown ecological traumas and losses in the natural world that are yet to come for Indigenous communities, "including the loss of species, ecosystems, and meaningful landscapes due to acute or chronic environmental change."[36] To address the ecological grief of peoples who retain close living and working relationships to the natural environment, we must expand our notions of trauma, loss, and repair beyond the singular event of violence and the singular figure of the human so as to consider the wholescale losses of culture, livelihoods, ways of life, and other nonhuman species and creatures inhabiting the natural world.[37]

In these ecological zones of sacrifice and suffering, the Sahtu Dene's apology not only exposes the political limits of the human and human rights but also the differential distribution of slow violence and ecological grief, their long-lasting biopolitical effects on humans and nonhumans alike, on forsaken communities abandoned by the sovereign state, and on bad objects deemed beyond political recognition, liberal representation, and the politics of repair. Absolute apology brings the Sahtu Dene together with other Indigenous communities suffering the toxic effects of uranium mining. It also unknowingly aligns them with the victims of atomic devastation in Hiroshima and Nagasaki, creating in a flash a bond among unanticipated communities, creatures, and things.

With the Sahtu Dene's pilgrimage to Hiroshima, the beginning meets the end. The first peoples of the Americas who transported the radioactive ore for the creation of the atomic bombs come together with the first communities in Asia on whom these weapons of mass destruction descended. Let us turn, then, to *A Village of Widows* to explore this surprising encounter.

A Village of Widows

Peter Blow's 1999 documentary film *A Village of Widows* opens with "another burial in Deline." In a somber voice-over that matches a gray and sunless day, the narrator tells us that in the Slavey language of the Sahtu Dene, Deline means "where the river flows." He immediately adds that the town is now known as "the Village of Widows." As a bell tolls over the cries of weeping mourners, and as a solitary dog howls into the wind, we hear the voice of Shirley Baton, a parent in the Deline community. As her visage slowly fills the screen, she tells us an unbearable story of death and disease: "My dad died of cancer, my aunt died of cancer, my grandmother died of cancer, my mother is suffering because of her sickness. And like what about my children? Cuz I still believe that once there is something in a person's genes, it carries on."[38]

Throughout Blow's moving and often enraging documentary, we are presented with similar stories of trauma, suffering, and loss, with a disturbing history of Dene locals who labored for decades at the behest of white strangers from the south, helping to extract and transport a deadly ore with no knowledge of its dangers, no insight on its use, and no awareness of its final destination. George Blondin, a Sahtu Dene elder, recounts,

> I had seven children and four of them died, plus my wife. And they all died of cancer. And I have spent fifteen years right in front of Echo Bay Mine. And I had my fish net right in front. And my dad lost three children. That's only two families, my dad and me. And almost all the families from Great Bear Lake that have lived close to Echo Bay Mine lost half of their children. And you don't have to be very educated to know

the possible effects from uranium that they threw in the water and that affects our life.[39]

These testimonies reveal that Deline could have as easily been named "the Village of the Dead." The radiation that ticks in the collective body of the community spreads without regard for gender or generation. In the aftermath, we are told by Cindy Kenny-Gilday, the chair of the Deline Uranium Committee, it produces a spoiled and wounded kinship. The passing of tradition from one generation to the next is shattered, marked by the chronic pain of cancer as well as the agony of knowing, the tense anticipation of a bleak future of unknown disasters still to come. These are the multiple and overlapping temporalities of knowledge, pain, and trauma the Dene must face, as the continuity of kinship is displaced by sickness in hindsight and illness in foresight.[40]

The first third of Blow's documentary narrates similar stories of family devastation, one involving the steady poisoning of the humans, the land, and the ecosystem. It was only in the 1990s, decades after the shuttering of uranium mining at Echo Bay, that the Canadian government revealed that 1.7 million tons of radioactive tailings (mine waste) remain unburied and scattered throughout the region, contaminating the water, soil, and air. A million tons of tailings alone lie at the bottom of Great Bear Lake—the ninth-largest freshwater lake in the world—from which the Sahtu Dene, animals, and plants draw their sustenance.

For the Sahtu Dene, these tailings are "found objects" that have served not only as contaminated "toys" with which their children once played but also, in the case of the Navajo in the United States, as construction materials for the very houses that sheltered them.[41] Unlike humans, who now understand the dangers of contaminated hot spots, the caribou, fish, and other wildlife migrate across these toxic areas without restriction, feeding off the lichen and drinking from the irradiated lake. Like the wildlife, the Dene must travel extensively on a seasonal and rotational basis around Great Bear Lake to hunt, fish, and gather food for consumption. "People are literally afraid of their own sustenance," Kenny-Gilday flatly states, commenting on the community's subsistence livelihood. "There is fear of their own lands. They don't know what the contamination levels are. And that's a bad way to live."[42]

A Village of Widows narrates this fifty-plus-year history of the "highway of the atom": the discovery of uranium at Port Radium; the transportation of radioactive ore along a 2,100-kilometer stretch of rivers, rapids, and portages; the construction of the world's first atomic bomb used on a human population; and the manifold aftereffects of that catastrophic nuclear event across time and space.

In the middle third of Blow's documentary, we see the Deline Uranium Committee, a group of elders and leaders in the community, travel to the nation's capital in Ottawa to petition the Canadian government for acknowledgment and redress. The committee holds an unprecedented meeting with three high-level cabinet ministers from the Departments of Indian and Northern Affairs, Public Health, and Energy and Mines.

In response to a 1994 Canadian government report identifying contaminated hot spots across their territory, the Deline Uranium Committee calls for recognition and relief. Although there are numerous expressions of sympathy all around, their demands remain unheeded by the state officials. To this day, there is little progress on the fourteen points for redress of health and environmental concerns that the committee presented to the ministers. The Canadian government discounts and dismisses both written requests and oral histories of Dene suffering and trauma. At one fraught point during their tense encounter—in a meeting room adorned with Aboriginal artworks—Andrew Orkin, a lawyer for the committee, bluntly tells his delegation, "They're trying to get you so tired and so unhappy and so frustrated that you just go back and you just give up."[43]

This meeting in Ottawa with state ministers represents a political scene of sovereign denial against which we must analyze the Sahtu Dene's journey to Hiroshima to apologize to the victims of the atomic bombing. Theirs is an apology all the more remarkable given the Canadian government's adamant refusal to acknowledge the toxic devastation and to respond to the ecological grief caused by uranium mining at Port Radium. Further, their act is all the more striking given the postwar Japanese government's reluctance to apologize for its crimes against humanity throughout East and Southeast Asia, along with the US government's refusal to apologize for the total destruction and death wreaked by the atomic bombings of Japan.

Recently declassified documents from the US Atomic Energy Commission reveal that both the US and Canadian governments were already aware of the dangers of uranium mining and radon gas exposure in the early 1930s.[44] Although the US government concluded that "an insufficient effort was made by the federal government to mitigate the hazard to uranium miners," leading the US Congress to pass the Radiation Exposure Compensation Act of 1990, Canada persists in its failure to address or redress the harms of its nuclear legacy.[45] In fact, a 2005 report commissioned by the Canadian government at a cost of $5 million attributed the exorbitant death rate among the Sahtu Dene to high incidences of alcoholism and tobacco use in the community, stating that there is no conclusive evidence that radiation exposure and contamination at Echo Bay were the source of the cancers that ravage the community.[46]

Significantly, mining is the largest industry in Canada, and historically uranium has been the country's largest and most profitable mineral export. Approximately 75 percent of the world's mining companies today are registered in Canada, although much of the actual mineral extraction operates in regions of the Global South.[47] Settler colonial states such as Canada fear that apology and reparations would entail recognizing Indigenous self-determination and necessitate the return of dispossessed lands to First Nation claimants. Simultaneously, it would require the abrogation of property rights, thereby curtailing economic benefits for those who have long profited off the stolen land in mining as well as in the agriculture, oil, and timber industries.

If reparations in the postwar period are thus framed by the state's sovereign authority—or, more accurately in this instance, sovereign refusal—to acknowledge and redress a long history of colonial injury for which it is directly responsible, then the Sahtu Dene's startling apology to the Hiroshima *hibakusha* (survivors of the atomic bomb) subverts and bypasses this sovereign mandate. Indeed, from a different perspective, their direct address to the survivors of the atomic bomb seizes this sovereign prerogative and, in so doing, reworks both its legal assumptions and its ethical failings across political and psychic divides.

Addressing the purpose of the Deline Uranium Committee's pilgrimage to Hiroshima, Kenny-Gilday states, "It's a justice issue for them [the committee]—on many, many, many levels. They have to make amends of some kind. So they have to go to the surviving relatives of the Japanese people and say, 'This is the way it happened.' And in telling that story they heal themselves."[48] The deliberate indifference of the sovereign state to acknowledge and repair that for which it is directly responsible meets its ethical rejoinder in the Deline Uranium Committee's 1997 journey to Japan. Their act of performative address, of response and responsibility—their attempt to give an account of themselves and the role they played in the atomic disaster—is an attempt to convey their sorrow and respect, in Modeste's words, as well as an attempt to heal the Deline community, in Kenny-Gilday's estimation. It thus returns us to the enduring dilemma explored in chapter 1 of this book—namely, the problem of how to repair bad objects unrecognized and abandoned by the settler colonial state.

The final third of Blow's documentary focuses on the committee's visit to various sites of memory in Hiroshima—the Hiroshima Peace Park and Museum, a *Tōrō nagashi* (lantern floating) ceremony for the dead, and different *hibakusha* communities sickened by radiation poisoning—and it is to this final part of Blow's documentary that I now turn.

Absolute Apology, Absolute Forgiveness

In "On Cosmopolitanism and Forgiveness," Derrida offers a counterintuitive notion of forgiveness, suggesting that the only thing worth forgiving is the unforgiveable. The unforgiveable is the "only thing that calls for forgiveness," he emphasizes. "Forgiveness is not, it *should not be*, normal, normative, normalizing. . . . It *should* remain exceptional and extraordinary, in the face of the impossible: as if it interrupted the ordinary course of historical temporality."[49]

On the one hand, Derrida suggests that what can be forgiven is invariably tied to political calculation, aligned with a set of intended political consequences and pragmatic effects. For example, a perpetrator apologizes, and a victim, in the face of this performative demand, is burdened with the imperative to forgive. Thus, the pair is inserted into a dialectic of apology and forgiveness that ultimately serves to underwrite liberal narratives of reparations, reconciliation, progress, and justice. "The language of forgiveness, at the service of determined finalities," Derrida observes, is "anything but pure and disinterested. As always in the field of politics."[50]

Returning to my analysis of Locke in chapter 1, we might understand apology and forgiveness as locked in a dialectic of political calculation and calculability that ultimately justifies the redistribution of property and life from the savage Indian to the European settler. In this context, apology functions more as apologia—a defense for the past excesses of the colonizer's violation of the Native.[51] It operates as an act of social rehabilitation for the colonizer that, as the Sahtu Dene's predicament underscores, privileges the worldview and interests of the settler over the welfare of the Native precisely through the demand for forgiveness.

As we discussed in relation to Klein, this dialectic also underwrites the settler's appropriation of the psychic suffering of Indigenous others. Not unlike the sympathetic expressions of the Canadian ministers in Ottawa to members of the Deline Uranium Committee, and similar to Chief Justice Marshall's perspective in his *Cherokee Nation v. State of Georgia* ruling, Native suffering is recognized precisely to be disregarded: "If courts were permitted to indulge their sympathies," Marshall admits, "a case better calculated to excite them can scarcely be imagined."[52] Yet the law's sympathies are severely limited.

Apology as apologia thus becomes a political alibi for a long history of colonial expansion and Indigenous dispossession, rather than an act of sincere contrition, of response and responsibility toward a violated other. In the process, it cannot establish or sustain a relational tie to the Native other. It cannot constitute the Indian as a good object worthy of recognition or repair. As either a ritualized political demand or a psychic act meant to appropriate the suffering of others, this strategic deployment of apology and forgiveness relieves the settler of both social responsibility and psychic guilt. In the same breath, it underwrites a process of political containment grounded in a narrative of historical progress that ultimately affirms the sovereignty of the settler state and its privileged citizen-subject.[53] In this instance, apology and forgiveness are akin to reparation as a noun, as an event that consigns a history of colonial violence and trauma into the past.

On the other hand, to forgive the unforgivable—to forgive in the absence of apology—shatters such liberal narratives of completion. Offered unconditionally, absolute forgiveness neither demands apology nor solicits repair. It is given without obligation or expectation, extended in the absence of reciprocity or exchange. As Derrida explains, absolute forgiveness is not "normal, normative, normalizing." By dissociating apology from forgiveness and victim from perpetrator, the calculability of politics thus cleaves to the incalculability of justice. In their wake, dialectics of sovereign authority, power, and will are short-circuited and shattered, demanding a reckoning with a state of exception that excludes the dispossessed Native from its liberal norms and values.[54]

In dreaming of the "purity" of a forgiveness that is "unconditional but without sovereignty," Derrida concedes that absolute forgiveness is "mad": it is a "madness of the impossible. . . . It is even, perhaps, the only thing that arrives, that surprises, like a revolution, the ordinary course of history, politics, and law. Because that means that it remains heterogeneous to the order of politics or of the juridical as they are ordinarily understood."[55] As absolute forgiveness marks a gap between law and justice, and between political calculation and ethical responsibility, it becomes more akin to reparation as a verb rather than a noun. It keeps open a space for new social relations to emerge, for unanticipated atrocities to be apprehended, and for future claims to be made by bad objects forsworn by the sovereign state.

If, as Derrida suggests, absolute forgiveness demands forgiving that which is unforgivable, I propose that absolute apology involves apologizing for that which you are not quite responsible. The Sahtu Dene's surprising act to the victims of Hiroshima interrupts, like uranium itself, the ordinary course of historical temporality. Dispossessed of their land, targeted by enormous state violence and neglect, and suffering from unspeakable personal losses and long-term environmental devastation, the Sahtu Dene nonetheless apologize. In the absence of political coercion or demand, and unaware that the yellow rock they helped to mine would be deployed for the massacre of countless victims in a distant land, the Sahtu Dene voluntarily assume the mantle of perpetrator. They take responsibility for their role in the atomic disaster, for a violence and misery that claims them as much as any other. What is at stake when those who might be considered the most vulnerable—the most precarious and victimized by the settler state—nonetheless assume the mantle of perpetrator?

The Sahtu Dene, by cleaving to the Japanese victims of the atomic bombing, assent to a notion of history as being "implicated in each other's traumas," returning to Cathy Caruth's trenchant formulation.[56] Exceptional and extraordinary, and in the face of the politically impossible, their surprising act blurs the line between victim and perpetrator itself. Indeed, it illustrates how one might be victim and perpetrator at once.[57] The ethical stakes of claiming such a subject position are high, for the Sahtu Dene's absolute apology presents us with a different idea of responsibility and repair, one in which we must take account of, as well as respond to, events beyond clear notions of agency and will, linear concepts of cause and effect, and even the immediate consequences and results of decisions clearly made by others.[58]

If responsibility emerges, as Emmanuel Levinas argues, as a consequence of being subject to the unwilled address of the other, the Sahtu Dene's response to the unanticipated call of the *hibakusha* demonstrates how violence and accountability are shorn from conventional politics of reparation and reconciliation.[59] The group's pilgrimage to Hiroshima loosens the dialectic of apology and forgiveness—the absolute distinction between victims and perpetrators—on which legal claims for reparations and human rights have been predicated. After all, legal claims for reparations and human rights are based on trauma not as a shared phenomenon in which we are all implicated, but, rather, as a psychic injury inflicted by one group on another in the liberal politics of victimhood and recognition. If trauma is a thing to be monopolized in conventional understandings of human rights violations, the Sahtu Dene's gesture underscores a different notion of repair that encompasses not ownership but, in Peter van Wyck's words, the "infinite character of responsibility."[60]

The Sahtu Dene's surprising act of apology without expectations of forgiveness is not only separated from calculated circuits of reciprocity and exchange—from the calculability of politics and the juridical as they are ordinarily understood—but also antithetical to absolute sovereignty of the state. It thus offers one way of approaching Derrida's dream of the "unconditional but without sovereignty." By acknowledging not the sovereign inviolability of the self-determined subject but a common vulnerability and susceptibility that binds together all creatures and things, it leaves us open to the unwilled address of an injured and unknown others. In the age of the Anthropocene and in the time of nuclear fallout, this profound act of compassion—*compassioun* in

Latin, meaning "a suffering with another"—exemplifies a willingness to become "undone in relation to others" and thereby constitutes a "chance of becoming human."[61] In the final analysis, it illustrates how we might begin to repair bad objects that remain to be seen.

If the Sahtu Dene's absolute apology points us toward an ethical ideal at the very limits of the political, indexing "a madness of the impossible," their "mad" act creates a provisional bond between the Sahtu Dene and the Hiroshima *hibakusha* that is neither absolute nor unconditional. We need not idealize or romanticize the Sahtu Dene's actions. Indeed, we learn in a follow-up documentary to *A Village of Widows* titled *Somba Ke: The Money Place* that, faced with no other means of subsistence, a younger generation of Sahtu Dene leaders have agreed to renew territorial leases at Echo Bay, opening up the ravaged area once again for uranium mining in a recursive cycle of colonial violence.[62] Absolute apology remains contingent on the absolute vicissitudes of history, configuring reparation not as a singular event but as an ongoing and unfinished process.

In the concluding segment of *A Village of Widows*, there are two back-to-back scenes bearing witness to the contingencies of absolute apology. In the first scene, Kenny-Gilday addresses the Hiroshima Peace Committee, a nongovernmental organization advocating for the abolition of atomic weapons and nuclear energy. She tells her Japanese audience,

> It is not only as an aboriginal person, a Dene from Deline, I take on a personal responsibility for what's happened here but also what is present now. Because it was the Canadian government that helped contribute to what we are now facing in India and Pakistan. And when I was a young woman in university, I protested against atomic bomb explosions testing off the coast of Alaska because I never ever wanted to see my daughter face this day. But she is now. So it's with gratefulness to all of you for holding up the shining light on a global level for the rest of us to follow, and I hope that this first visit will become a pilgrimage for peace from our people, and that we will continue working together and praying together for peace.[63]

Referring to India and Pakistan's refusal to participate in the nonproliferation treaty, as well as the continued buildup of nuclear weapons that will invariably harm future generations, Kenny-Gilday speaks as both confessor and critic. Ironically, while the pain and suffering of the Sahtu Dene goes unacknowledged by the Canadian government, Kenny-Gilday recognizes the historical role that *both* groups—the Canadian government as well as the Dene—have played in the current geopolitical tensions dividing the South Asian continent. She thereby takes responsibility for not only what has happened in the past (the atomic bombings of Japan) but also what is present now (continued nuclear testing and the atomic standoff between contending nations). Notably, Kenny-Gilday takes responsibility both as a representative of the Dene peoples who seeks, in her own words, to give an account of "the way it happened" and as a critic of the sovereign Canadian state, of which the Dene are simultaneously a part and apart.

These "nested sovereignties"—in the words of Kahnawake Mohawk scholar Audre Simpson—index alternative economies of thought and practice. They illustrate that within Canada there exist other political orders, including that of the Sahtu Dene, which interrupt the conventional mandates of the sovereign state and its liberal politics of recognition. In this regard, the unconditional sovereignty of the state and its sanctioned citizen-subject as the universal bearer of liberal rights and representation remains antipathetic to the absolute apology of the Sahtu Dene, one insisting on the integrity of the ethical mandates guiding their actions in relation to others.

The Dene's absolute apology, offered outside the liberal politics of calculation and calculability, puts considerable pressure on Western models of sovereignty, citizenship, temporality, and progress that have subjected Indigenous peoples to secular as well as religious processes of colonization and conversion. There is widespread debate in Indigenous studies among scholars on the political uses and abuses of sovereignty.[64] A lodestar around which numerous Indigenous political movements for decolonization have been organized, sovereignty has served as a vitally important rallying cry for social justice movements, the enforcement of treaty rights, and the reclamation of unceded land, and as a defense against neoliberal enclosure and extractive economies, including uranium mining, oil and gas pipelines, and the damming of rivers and lakes.[65]

This chapter is not the occasion to rehearse comprehensive debates on sovereignty in Indigenous studies.[66] Along with Dene scholar Glen Sean Coulthard, I draw attention instead to how alternative worldviews "deeply inform and sustain Indigenous modes of thought and behavior that harbor profound insights into the maintenance of relationships within and between human beings and the natural world built on principles of reciprocity, nonexploitation and respectful coexistence."[67] As a system of reciprocal relations and obligations, the land practices of Indigenous peoples, what Coulthard describes as "landed normativity," provide "invaluable glimpses into the ethical practices and preconditions required for the construction of a more just and sustainable world order."[68]

As Gina Bayha, a member of the Deline Uranium Committee, states, "To us, the land and the resources and everything is very sacred because of the fact that we rely on it to continue to live. And that very source is what actually caused damage to another people." Explaining the motivation behind the committee's pilgrimage to Hiroshima to apologize and make amends for the destruction that came from their lands, Bayha notes that the Dene need "to acknowledge that this actually happened. Yet at the same time, we acknowledge that something as sacred as that that came from the land could be just as harmful."[69] She thus indexes the ethical ties and obligations that bind together land, animals, plants, and lakes as well as peoples across the Transpacific.

In *The Art of Not Being Governed*, the anthropologist and political scientist James C. Scott investigates populations outside formal zones of state sovereignty. These hill tribes occupying the Zomia uplands of Southeast Asia evade state-making projects that would seek to conscript, enslave, exploit, and tax them while embroiling their communities in warfare and conflict. Extending Scott's study, the Sahtu Dene's absolute apology, one emerging from their nested sovereignties, landed normativity, and alternative worldview, maps a different approach to the art of not being *fully* governed *within* rather than outside the sovereign state.[70] This *within* offers alternative tactics for coexistence with as well as beyond the sovereign settler state.

Directly following Kenny-Gilday's speech to the Hiroshima Peace Committee, we are presented with a second scene of absolute apology—a scene of surprise—that extends problems of self-determination and mutual reciprocity across different subaltern communities, spaces, and times. Here the Deline Uranium Committee visits a *hibakusha* hospital, but one that, the narrator explains, is racially segregated. The institution is a medical clinic specifically designated for Korean *hibakusha*, thirty thousand forced laborers exposed to radiation poisoning from the atomic bombing. This history of Japanese colonization and racial subordination is, as Lisa Yoneyama point outs, "virtually absent from past official representations of Hiroshima's atomic atrocity."[71]

Blondin tells his audience, who, like his own community in Deline, are suffering from their sickness, "We come here to learn. We heard about the problem of A-bomb—which people have really suffered. We as an Indian, we share your sorrow, our sorrow, and we share that together. I am part of you. Indian law goes like that. There's no stranger in the world. Everybody is your brother and sister, and as an Indian we love each other. So, therefore I love you people, and I see that you are my brother and sister. That is my general thinking as an elder."[72] This is a fascinating and exceptional scene—a mad scene, even—as it interrupts, like a revolution, the traditional order of things.

First, Blondin's address to these Korean *hibakushas* extends the act of compassion introduced by Kenny-Gilday: "We come here to learn. We heard about the problem of A-bomb—which people have really suffered. We as an Indian, we share your sorrow, our sorrow, and we share that together. I am part of you. Indian law goes like that." In the same breath that Blondin asserts a mutual kinship of shared suffering between the two groups, he also qualifies the sovereign state of exception founded on the constitutive exclusion of others. "There's no stranger in the world," Blondin asserts. "Everybody is your brother and sister, and as an Indian we love each other. So, therefore I love you people, and I see that you are my brother and sister." Through both compassion and the acknowledgment of the reciprocal relations among all creatures and things in intimate surroundings, Blondin's observations thus work to supplement the dominant narrative of a "nuclear universalism" emerging in the wake of atomic disaster.

Like Chakrabarty's universal concept of species being, we imagine atomic destruction today in the language of a nuclear universalism that threatens the existence of all humankind—every living creature and thing on planet Earth. In the process, it transforms, as Kenny-Gilday suggests in her speech to the Hiroshima Peace Committee, the atomic catastrophes in Japan into a universal symbol for global peace. Nuclear universalism has in fact become an idée fixe in Japanese national history that, while serving to bring disparate groups together under the sign of a threatened humanity, also serves to disavow a history of both Japanese colonialism and Western empire in the region whereby, in the words of anthropologist Marilyn Ivy, "victimization takes precedence over the guilt of atrocities committed."[73] This reification of Japanese victimhood in the wake of atomic devastation transforms the Japanese enemy and bad object—borrowing from the psychoanalytic terminology of the previous chapter—into Japanese friend and good object worthy of US rehabilitation and repair.

Taken together, these two scenes of Dene apology addressing two different groups in the *hibakusha* community raise an urgent question in the face of assertions of nuclear universalism: What does it mean for the Sahtu Dene to apologize to Japanese *hibakusha*—citizen-subjects of a colonial empire in Asia—and what does it mean for them to apologize to Korean *hibakusha*—colonized subjects forced into conscripted labor by the Japanese imperial war machine?[74] These 30,000 surviving victims of the atomic bombing are part of Zainichi communities, 700,000 Korean forced laborers brought to Japan under colonial rule between 1939 and 1945 to work in coal mines, munitions factories, and other dangerous wartime sites. (The term *Zainichi* denotes "a foreign citizen of another country temporarily residing in Japan.") Today Zainichi comprise the second-largest minority group in Japan, the majority of whom are permanent foreign residents rather than formal citizens of Japan. They remain second-class citizens, subject to carrying alien registration cards at all times, and subordinated by a long history of colonial domination and racial discrimination.

There is an important and telling acoustic detail in the soundtrack of this hospital scene: namely, the entire exchange between Blondin and the Korean *hibakusha* is mediated not in their native tongues—Slavey and Korean—but through the dominant language of the colonizers, which, while marking their linguistic compliance, cannot fully overdetermine their ethical will. That is, both Blondin and the Korean *hibakusha* are impelled to narrate their sufferings in English and Japanese, respectively, with two Japanese-English translators serving to negotiate this already highly mediated linguistic exchange. These multiple linguistic transformations not only particularize the dream of a universal humanity under atomic threat but also disrupt the collective bond of shared suffering between the Sahtu Dene and the Korean *hibakusha*.

To put it otherwise, this linguistic mediation opens onto a larger historical and political landscape of state violence and trauma configured by atomic disaster that brings together, even as it pulls apart, these two subaltern communities across the Transpacific. Here it might be useful to recall Thomas Jefferson's assertion in his *Notes on the State of Virginia* that Indians in the Americas, having migrated across the Bering Strait, are not Indigenous to the New World but are in fact Asian. As such, Jefferson reasoned, they have no greater claims on the land of the Americas than colonial settlers from Europe do. This particular linking of the figures of the Indian and the Asian, as Chickasaw scholar Jodi A. Byrd points out, configures Native Americans as the first Asians in the United States, the first wave of Yellow Peril in the Americas, and thereby a community with no rights to Native title.[75]

Even as Jefferson's analysis warns us of the political pitfalls of any gestures toward a desired universalism, it cautions us to remain vigilant to how the linking of dispossessed and colonized victims of atomic disaster has been hijacked by liberal political representations of race. As Byrd warns us, "The conflation of racialization into colonization and Indigeneity into racial categories dependent upon blood logics underwrites the institutions of settler colonialism when they proffer assimilation into the colonizing nation as reparation for genocide and theft of lands and nations."[76] The erasure of Indigeneity is the condition of possibility for the nation-state and liberal racial representation to emerge.

The linguistic perils of Blondin's exchange with the Korean *hibakusha* focus our critical attention on the necessity of accounting not only for Western settler colonialism across the Transpacific but also for Japanese empire throughout East and Southeast Asia. It forces us to pay heightened attention to the overlapping accounts of dispossession and empire across the Americas and Asia that simultaneously bring together as they separate the figures of the Indian and the Asian. Framed by an uneven history of trauma and violence, and in the face of the persistent limits of reparations and human rights, the contingent bonds that Kenny-Gilday's and Blondin's absolute apologies establish with their counterparts in Hiroshima are necessarily fleeting. Yet in their contingency they impart an invaluable lesson for a different approach to reparations and the human, and for alternate models of political and psychic repair beyond the nation-state.

Ultimately, if injury is not reparable, and only relationships are, then the Sahtu Dene's unanticipated attachments create a new relationality precisely through and despite a history of injury. These attachments sustain new forms of community and solidarity that did not exist before. They honor the mutual vulnerability and unwilled susceptibility that form their distinct relations in excess of sovereignty and citizenship. "This bond cannot be contained within traditional concepts of community, obligation or responsibility," in Derrida's words. It "is a protest against citizenship, a protest against membership of a political configuration as such."[77]

As Blondin concludes his remarks to the Korean *hibakusha* in this hospital scene, which are translated into Japanese by a young woman nearby (Translator 1), he is greeted by a round of hearty applause and *arigatos* (thank-yous). He is then offered a firm handshake by one of the Korean *hibakusha* who have gathered on the hospital beds around him. This man speaks, and the second translator in the room interprets:

> TRANSLATOR 2: He says, "We have the same skin color" and, yeah, so we feel . . .
>
> BLONDIN: Oh, not quite the same!
>
> TRANSLATOR 2: Not quite the same, but we look alike, so we feel . . .
>
> TRANSLATOR 1: He thought you were Japanese when he first saw you!
>
> BLONDIN: I was thinking of to get married in Japan. But maybe I'm no good-looking. Nobody look at me . . .

This scene of heightened feeling deserves attention. Mistaken for the Japanese colonizer, Blondin's rejoinder—"Oh, not quite the same," along with his wishes "to get married in Japan"—fills the room with collective laughter.

Despite the misrecognition of their significant national and racial differences —confusions between victim and perpetrator, colonizer and colonized, the Indigenous and the racialized, those dispossessed of their land and those exploited for their labor—an atmosphere of palpable conviviality descends on the group. This affective tie, I suggest, brings the Sahtu Dene and Korean *hibakusha* together despite their considerable differences, and in a manner distinct from the ways their propinquity is instrumentalized by the sovereign liberal state. This convivial tie—in Latin, *convivial* means "living with"—transforms their mutual but distinct histories of collective suffering, of dispossession and colonization across the Americas and Asia, into a new mode of becoming.

Most significantly, to return to my earlier discussion of psychic genealogies of reparation, this convivial tie illustrates how bad objects abandoned by the sovereign state might nonetheless acknowledge and repair *each another*. This living together *with* one another is enabled not through gestures of sympathy that seek to colonize the suffering of the other but, rather, through acts of profound compassion and mutual recognition. Previously unknown to one another, the two group's contingent bond is established precisely through the acknowledgment of an unwilled address and by an ethical response to unintended harm. Thus, the Sahtu Dene's absolute apology moves us beyond the insuperable dialectics of aggression and defense, guilt and blamelessness, perpetrator and victim, apology and forgiveness.

To reframe these observations more directly in line with my analysis of colonial object relations in chapter 1 of this book, what does it mean for a *bad* object (the dispossessed Sahtu Dene) to redress and repair a *good* object (the sanctioned Japanese citizen-subject, the Japanese *hibakusha*)? And what does it mean for a *bad* object to redress and repair *another bad* object (the colonized Korean *hibakusha*)? These questions return us to Derrida's notions of absolute forgiveness and the asymmetrical power relations that are constituted by the liberal state and reflected in its politics of apology and forgiveness.

On the one hand, while the Sahtu Dene's gesture to the Japanese *hibakusha* might be seen as outside conventional economies of reciprocity and exchange, it nonetheless works to reinforce the sovereignty of the Japanese nation-state and the dominant narrative of victimhood and nuclear universalism resolutely attached to Hiroshima memories. Yoneyama observes that "whether in mainstream [Japanese] national historiography, which remembers Hiroshima's atomic bombing as victimization experienced by the Japanese collectivity, or in the equally persuasive, more universalistic narrative on the bombing that records it as having been an unprecedented event in the history of humanity, Hiroshima memories have been predicated on the grave obfuscation of the prewar

Japanese Empire, its colonial practices, and their consequences."[78] For better or worse, the Sahtu Dene's mad gesture to the Japanese *hibakusha* is conscripted into this dominant political narrative.

On the other hand, the Sahtu Dene's apology to the Korean *hibakusha* is an unsolicited response that provides a road map beyond the sovereign authority of the state—its failure to acknowledge and to redress the injuries for which it is directly responsible: injuries against the dispossessed Sahtu Dene in Canada, the colonized Korean *hibakusha* in Japan, or the many other victims indiscriminately annihilated by US bombs in Japan and elsewhere. By illustrating how bad objects neglected and abandoned by the settler state can recognize and repair one another, this quiet scene of surprise and mirth points us to the possibility of what I might call "affective reparation." That is, it asks us to recognize the alternative ways in which an aesthetics of redress can take hold precisely through psychic processes of reparation embodied by the Sahtu Dene's ethical embrace of response, responsibility, compassion, and conviviality. In short, it illustrates how an affective politics of care toward an unwilled other can, indeed, take a political form.

Chief Justice Marshall's opinion in *Cherokee Nation* declares that Cherokees have no standing in court to sue the state of Georgia, that they have no legal ground to stand on. Byrd argues that histories of Native dispossession in the Americas have rendered the figure of the Indian groundless; indeed, the Indian becomes the very ground on which liberal representation and racial figuration are constructed.[79] In the wake of dispossession, liberalism cannot account for, address, or repair those bad objects that remain groundless, that are without ground, that have no legal standing in a long and unrequited history of appeals to the law for justice. In contrast, the unexpected mirth engendered by this scene illustrates how affect might come to offer a different means for racial figuration, a new aesthetics for dispossession, a different route for the representation of bad objects without standing, that are groundless, that are outside the boundaries of the law and still searching for a political form.

I have examined throughout this book how conventional political approaches to reparations and the human have created a host of devalued, disparaged, and ghostly figures: Indigenous others, Korean *hibakusha*, comfort women, war orphans, and refugees. These figures are incapable of being politically represented or repaired—of being recognized as human. The problem as well as promise of racial reparations demands a reworking of the unequal distribution of trauma and loss that proleptically creates the privileged figure of the violated human being worthy of attention and redress. The unacknowledged violence across the Americas and Asia, human and nonhuman, raises the query of how to rearrange our political and psychic desires beyond sovereignty, trauma, and repair in a way that would transform bad objects unworthy of care and repair—of being human—into good objects endowed with value. Even more, the fact of fallout and the prospect of global environmental disaster demand that we also move beyond the dialectics of apology and forgiveness, of good and bad, perpetrator and victim. That is, they require that we move beyond history as only human history and human conflict.[80]

In the final analysis, absolute apology exposes the political failures of both the sovereign state and human rights to protect the sanctity of all human and nonhuman life in the age of the Anthropocene. At the same time, it proposes a way beyond these political and psychic predicaments for a different model of reparations and the human to emerge. Law penalizes individuals for what they should not do rather than guiding them to care for each other and to create bonds of mutual compassion and conviviality. Law functions under the sign of the sovereign and the universal and therefore cannot, in the words of feminist theorist Carolyn D'Cruz, account for the singular and the specific, for "the other who slips beneath the calculability of law and cannot be subsumed by a general principle."[81] The Sahtu Dene's absolute apology is a performative act that unfolds a singular and specific, a contingent and unanticipated scene of care and repair, that directs us away from political calculation and calculability toward a noncoercive rearrangement of our projections and desires.[82] It dissociates the sovereign from the unconditional, the universal from the particular, the good object from the bad, and reparations from racial reparations. We repair only that which we care about, that which is valuable to us. "Nobody look at me," Blondin laments to his convivial audience. Instead, they look at each other.

Notes

PREFACE

1 *OED Online*, s.v. "human (*n*, 2a)," accessed July 25, 2024, https://www.oed.com/dictionary/human_adj?tab=meaning_and_use#1122977.

INTRODUCTION

1 Arendt, *The Origins of Totalitarianism*, 298, 299.
2 Minow, "Breaking the Cycles of Hatred," 14.
3 Anghie, "Politic, Cautious, and Meticulous," 66.
4 See Fanon, *Black Skin, White Masks* and *The Wretched of the Earth*; Wynter, "Unsettling the Coloniality of Being/Power/Truth/Freedom"; Spillers, "Mama's Baby, Papa's Maybe"; Weheliye, *Habeas Viscus*; and Jackson, *Becoming Human*.
5 Fanon, *The Wretched of the Earth*, 1.
6 Spelman, *Repair*, 8. Spelman writes, "For though we do not repair everything we value, we would not repair things unless they were in some sense valuable to us, and how they matter to us shows up in the form of repair we undertake" (8).
7 Lowe, "The Intimacies of Four Continents," 208, 206.
8 See Mills, *The Racial Contract*.
9 See Brown, *States of Injury*.
10 For an account of the "human being after genocide" and in the face of unprecedented postwar biopolitical weapons of mass destruction, see Wald, "Exquisite Fragility."
11 According to legal historian Samuel Moyn, the postwar ascendance of human rights came in fits and starts. A discourse of human rights appeared in the wake of the Holocaust, including the UN's "Universal Declaration of Human Rights" in 1948. However, it was not until the mid-1970s that contemporary notions of human rights as they are conceived today were finally established as the "last utopia" in the wake of the failures of other internationalist movements. See Moyn, *The Last Utopia*.

12 Douzinas, *Human Rights and Empire*, 33.

13 There is considerable debate on the use of the term *comfort women*. Although it is the nomenclature that has gained the most purchase in international human rights discourses since the 1990s, feminist scholars have argued that the term *military sexual servitude* or *militarized sexual labor* is more accurate. See Kang, *The Traffic in Asian Women*; and Moon, "Military Prostitution and the U.S. Military in Asia."

14 Civil Liberties Act of 1988, H.R. 422, 100th Cong. (1988).

15 See Nguyen, *The Gift of Freedom*.

16 See J. W. Scott, *On the Judgment of History*.

17 Butler, *Giving an Account of Oneself*, 44.

18 See Rothberg, *The Implicated Subject*.

19 Caruth, *Unclaimed Experience*, 24.

20 Wynter, "Unsettling the Coloniality of Being/Power/Truth/Freedom," 268.

21 See Latour, *We Have Never Been Modern*.

22 See Pietz, "The 'Post-colonialism' of Cold War Discourse."

23 See Hopgood, *The Endtimes of Human Rights*; Perugini and Gordon, *The Human Right to Dominate*; and Williams, *The Divided World*.

24 See Trouillet, *Silencing the Past*; and Dubois, *Haiti*. Haiti agreed to pay in 1825 150 million francs—the equivalent of $21 billion in 2020 terms—in order to gain the political recognition that would end the fledgling nation's diplomatic and economic isolation. In 1828, the sum was lowered somewhat, but Haiti made debt and interest payments from 1825 through 1947. This onerous debt continues to devastate Haiti's political stability, economic prosperity, and social welfare.

1. BEYOND REPAIR

1 Freud, *Beyond the Pleasure Principle*.

2 Freud formally gathers these inexplicable psychic complexes under the name "death drive" in 1920. However, as early as 1915, as we witness in *Thoughts for the Times on War and Death*, Freud was already contemplating what we might describe as the socialization of the death drive. Writing directly amid the engulfing violence of World War I, Freud argues that Western civilization has lost its ethical bearing, killing with impunity, while refusing to be haunted by the death of the other it has directly brought about through war, violence, and weapons of mass destruction.

3 Freud, *Thoughts for the Times on War and Death*, 277–78. Chillingly, Freud continues in regard to Germany:

> Moreover, it has brought to light an almost incredible phenomenon: the civilized nations know and understand one another so little that one can turn against the other with hate and loathing. Indeed, one of the great civilized nations is so universally unpopular that the attempt can actually be made to exclude it from the civilized community as "barbaric," although it has long proved its fitness by the magnificent contributions to that community which it has made. We live in hopes that the pages of an impartial history will prove that that nation, in whose language we write and for whose victory our dear ones are fighting, has been precisely

the one which has least transgressed the laws of civilization. But at such a time who dares to set himself up as judge in his own cause? (278)

All three of Freud's sons—Jean-Martin, Oliver, and Ernest—were in the Austrian military, although Oliver was rejected for military service until late in the war (1916), when he was called on to serve in a number of engineering projects on the Italian front. See Gay, *Freud*, 352–53.

4 Freud, *Thoughts for the Times on War and Death*, 276.
5 Arendt, *The Origins of Totalitarianism*, 299. See also Jefferies, *Contesting the German Empire*.
6 Notably, historians have attributed the rise of the Third Reich in part to the exorbitant reparations exacted from Germany after World War I. Following their defeat, Germany was forced to decolonize. As such, unfinished colonial ambition and violence in postcolonial Germany after the Treaty of Versailles was subsequently deflected inward toward (Eastern and Central) Europe as well as toward its own populace and persecuted minorities. See Poiger, *Jazz, Rock, and Rebels*; and Schumann, *Political Violence in the Weimar Republic, 1918–1933*.
7 In addition to work by Arendt, see Hull, *Absolute Destruction*; Lindqvist, *"Exterminate All the Brutes"*; Moses, *Colonialism and Genocide*; Olusoga and Erichsen, *The Kaiser's Holocaust*; Sarkin, *Germany's Genocide of the Herero*; Traverso, *The Origins of Nazi Violence*; and Weitz, *A Century of Genocide*. See also Gerwarth and Malinowski, "Hannah Arendt's Ghosts"; and Madley, "From Africa to Auschwitz."
8 See Césaire, *Discourse on Colonialism*; Fanon, *The Wretched of the Earth*; Mannoni, *Psychology of Colonization*; and Du Bois, *The World and Africa*.
9 See Sartre's foreword to Fanon's *The Wretched of the Earth*. I borrow these suggestive formulations from Gewarth and Malinowski, "Hannah Arendt's Ghosts." Arendt also uses this "boomerang" metaphor in *The Origins of Totalitarianism*.
10 See Sedgwick, *Touching Feeling*; and Berlant, *Cruel Optimism*.
11 Much of this scholarship has focused on nineteenth-century Europe, with the apotheosis of colonial enterprises, rather than on earlier periods in the seventeenth century. See, for example, Pitts, *A Turn to Empire*.
12 See Armitage, "John Locke, Carolina, and the 'Two Treatises of Government'"; Laslett, "Introduction"; Mehta, "Liberal Strategies of Exclusion"; and Tully, *An Approach to Political Philosophy*.
13 See, for example, Hunt, *Inventing Human Rights*.
14 See Tully, *An Approach to Political Philosophy*:

As secretary to Lord Shaftesbury, secretary of the Lord Proprietors of Carolina (1668–71), secretary to the Council of Trade and Plantations (1673–74), and member of the Board of Trade (1696–1700), Locke was one of the six or eight men who closely invigilated and helped to shape the old colonial system during the Restoration. He invested in the slave-trading Royal African Company (1671) and the Company of Merchant-Adventures to trade with the Bahamas (1672), and he was a Landgrave of the proprietary government of Carolina. His theoretical and policy-making writings on colonial affairs include the *Fundamental Constitutions of Carolina*

(1669), Carolina's agrarian laws (1672), a reform proposal for Virginia (1696), memoranda and policy recommendations for the boards of trade, covering all the colonies, histories of European exploration and settlement, and manuscripts on a wide range of topics concerning government and property in America. (140–41)

15 Armitage, "John Locke, Carolina, and the 'Two Treatises of Government,'" 603.
16 Locke, *Two Treatises of Government*, book 2, §27, 287; emphasis in original.
17 Locke, *Two Treatises of Government*, book 2, §27, 287–88.
18 Locke, *Two Treatises of Government*, book 2, §40, 296; emphasis in original.
19 Laslett, "Introduction," 103.
20 Locke, *Two Treatises of Government*, book 2, §49, 301; emphasis in original.
21 Locke, *Two Treatises of Government*, book 2, §34, 291; emphasis in original.
22 Tully, *An Approach to Political Philosophy*, 139.
23 Chakrabarty, *Provincializing Europe*, 8. For an excellent account of how Indigenous dispossession is based on a recursive colonial logic that produces what it presupposes—that is, the idea of property rights prior to land theft—see Nichols, *Theft Is Property!*
24 Locke, *Two Treatises of Government*, book 2, §41, 297.
25 Mehta, "Liberal Strategies of Exclusion," 448.
26 See Saldaña-Portillo, *Indian Given*.
27 In contrast to medieval historian Ernst Kantorowicz's notion of the king's "two bodies"—one corporeal and the other eternal—I argue that reparation's two bodies are based on the geographical displacement of the concept across the Transatlantic. Insofar as Locke also transposes the concept of the corporeal and eternal body onto the liberal individual, we might say that the European laboring body (corporeal) is brought together with the sovereign body (eternal) of the abstract individual. See Kantorowicz, *The King's Two Bodies*.
28 See Ames, *Lectures on Legal History and Miscellaneous Legal Essays*.
29 See Lazenby, *The First Punic War*.
30 Thucydides, *The History of the Peloponnesian War*, 402.
31 Locke, *Two Treatises of Government*, book 2, §182, 390; emphasis in original.
32 Locke, *Two Treatises of Government*, book 2, §182, 389.
33 See Schmitt, *The* Nomos *of the Earth*.
34 Ruskola, "Canton Is Not Boston," 862–63.
35 Anghie, "Francisco de Vitoria and the Colonial Origins of International Law," 322.
36 Anghie, "Francisco de Vitoria and the Colonial Origins of International Law," 328.
37 Locke, *Two Treatises of Government*, book 2, §130, 237; emphasis in original.
38 See Saldaña-Portillo, *Indian Given*.
39 Tully, *An Approach to Political Philosophy*, 147.
40 Laslett, "Introduction," 96.
41 Jackson, *Becoming Human*, 1, 2.
42 Tully, *An Approach to Political Philosophy*, 145. He observes, "The fact that the chapter is organized around a contrast between Europe, where appropriation without consent is not permitted because political societies exist, and America, where ap-

propriation without consent is permitted because it is in a state of nature, is rarely mentioned."

43 Tully, *An Approach to Political Philosophy*, 154–55. He writes:

> If he had recognized these forms of property, as Roger Williams and many others who signed treaties did, European settlement in America without consent would have been illegitimate by his own criteria of enough and as good. In addition, Locke has a further reason not to recognize the traditional property of the Amerindians. The argument for dispossession by agricultural improvement was often supplemented by the natural law argument for just conquest if the native people resisted. But, in Locke's theory of conquest (written for another purpose) the conqueror has no title to the property of the vanquished (180, 184). The conqueror has no right "to dispossess the Posterity of the Vanquished, and turn them out of their Inheritance, which ought to be the Possession of them and their Descendents to all Generations." Therefore, if the Amerindians had property in their traditional land conquest would not confer title over it. However, as Locke repeats twice in this section, in the case of conquest over a people in the state of nature, "where there . . . [is] more *Land*, than the Inhabitants possess, and make use of," the conqueror, like "any one[,] has liberty to make use of the waste" (184); thereby bringing his theories of conquest and appropriation into harmony.

44 On the inclusion of the Indian for exclusion, see O'Brien, *Firsting and Lasting*; for the figuration of the liberal human subject through the grounding of the Indian, see Byrd, *The Transit of Empire*.
45 The Cherokee Nation v. the State of Georgia, 30 U.S. 1 (1831), 20.
46 Barker, "For Whom Sovereignty Matters," 16, 17; emphasis in original.
47 *Cherokee Nation v. State of Georgia*, 15.
48 *Cherokee Nation v. State of Georgia*, 15.
49 Barker, "For Whom Sovereignty Matters," 14, quoted in Byrd, *The Transit of Empire*, xxi.
50 See Bickel, *The Morality of Consent*, 53.
51 See Locke, *An Essay concerning Human Understanding*. First appearing in 1689, with a publication date of 1690, *An Essay concerning Human Understanding* was roughly contemporaneous with the publication of *The Second Treatise* (1689), although there is considerable debate about the dates when the *Two Treatises* were composed. See Armitage, "John Locke, Carolina, and the 'Two Treatises of Government,'" on the publication chronology. It is worth considering how the former might be approached as a psychic reflection of the colonial mandates of the sovereign European state now embedded in the consciousness of the sovereign European individual.
52 Caruth, *Empirical Truths and Critical Fictions*, 43.
53 See Cohen, *A Body Worth Defending*.
54 Freud, *Beyond the Pleasure Principle*, 14.
55 I borrow this formulation from Rose, *Why War?*
56 Freud, *Beyond the Pleasure Principle*, 14–15.
57 Freud, *Beyond the Pleasure Principle*, 15.

58 Freud, *Beyond the Pleasure Principle*, 16.
59 Freud, *Beyond the Pleasure Principle*, 16.
60 More locally, the "controversial discussions" from January 1943 to May 1944 revealed a history of "total war" within the British psychoanalytic school. Klein and her writings on reparation were centrally embroiled in this war, in fractious disagreements on Freud's legacy in the United Kingdom, leading to the school's dissolution into three separate branches (Kleinian, Anna Freudian, and independent) that exist to this day. For a detailed history of this splitting, see Shapira, *The War Inside*.
61 Riviere, "On the Genesis of Psychical Conflict in Earliest Infancy," 416, quoted in Rose, *Why War?*, 166.
62 Rose, *Why War?*, 18. She writes, "But, the death drive, and hence the truth of war, operates, it has often been pointed out, as the speculative vanishing point of psychoanalytic theory, and even, more boldly, of the whole of scientific thought."
63 Phillips, "Editor's Note," 127.
64 Sedgwick, "Melanie Klein and the Difference Affect Makes," 629.
65 Sedgwick, *Touching Feeling*, 126.
66 Klein, "Love, Guilt and Reparation," 68.
67 Rose, *Why War?*, 166.
68 Klein, "Love, Guilt and Reparation," 68.
69 Klein, "The Psycho-analytic Play Technique," 48.
70 Butler, "Moral Sadism," 185.
71 Zaretsky, "Melanie Klein and the Emergence of Modern Personal Life," 39.
72 See Freud, *Mourning and Melancholia*.
73 Butler, "Moral Sadism," 184.
74 Butler, "Moral Sadism," 187.
75 On rudimentary knowledge, see Phillips, "Editor's Note," 12. On the child's fantasy of reparation versus the object's need for repair, see Laubender, "Beyond Repair," 57.
76 Butler, *Frames of War*, 25.
77 Butler, *Frames of War*, 46.
78 Butler, "Moral Sadism," 185.
79 For a sweeping survey and critical analysis of the scholarship on manic reparation, see Seitz, "A Wizard of Disquietude in Our Midst," 9. He writes that what makes reparation "manic" is "phantasies of omnipotence and autonomy—of infinite capacity to destroy and repair the object, sometimes including the phantasy of complete independence from an object, rendering repair unnecessary."
80 Klein, "Love, Guilt and Reparation," 68.
81 Butler, *Frames of War*, 177.
82 I draw this formulation from Rose, *Why War?*, 153.
83 Klein, "Love, Guilt and Reparation," 95.
84 Klein, "Love, Guilt and Reparation," 96.
85 Klein, "The Psychogenesis of Manic-Depressive State," 120.
86 Klein, "Love, Guilt and Reparation," 98.
87 Klein, "Love, Guilt and Reparation," 103.

88 Segel, *Introduction to the Work of Melanie Klein*, 35.
89 Klein, "Love, Guilt and Reparation," 97.
90 Klein, "Love, Guilt and Reparation," 103–4.
91 Klein, "Love, Guilt and Reparation," 104–5.
92 See Chambers-Letson, "Reparative Feminisms, Repairing Feminism," 175.
93 For an analysis of the problem of ambivalence in Klein's theory of reparation, see Stacey, "Wishing Away Ambivalence."
94 Klein, "Love, Guilt and Reparation," 91–92.
95 Butler, *Frames of War*, 177.
96 Musser, *Sensational Flesh*, 102. See Foucault, *The Care of the Self*, as well as his final set of lectures at the Collège de France, *Hermeneutics of the Subject*.
97 Butler, *Giving an Account of Oneself*, 99–100.
98 Rose, *Why War?*, 25.
99 Zaretsky, "Melanie Klein and the Emergence of Modern Personal Life," 38.

2. BEYOND TRAUMA

1 McCarthy, "The Hiroshima *New Yorker*," 367.
2 Arendt, *The Origins of Totalitarianism*, 299.
3 According to Alex Wellerstein in the *Bulletin of Atomic Scientists*, a low estimate of 110,000 and a high estimate of 210,000 deaths is most likely. Wellerstein notes that the number is "probably fundamentally unknowable. The indiscriminate damage inflicted upon the cities, coupled with the existing disruptions of the wartime Japanese home front, means that any precise reckoning is never going to be achieved." See Wellerstein, "Counting the Dead at Hiroshima and Nagasaki."
4 See Chang, *The Rape of Nanking*, 222. According to Chang, the German government had paid over $100 billion in reparations to date (in 1997). By contrast, its ally Japan, from the perspective of its wartime victims and former colonial subjects, has not properly been held to political or economic account for the horrors it perpetrated.
5 See Yoneyama, *Hiroshima Traces*, esp. 16–18. Yoneyama provides a detailed analysis of the epitaph debate concerning the inscription on the Memorial Cenotaph. See also the Wikipedia entry on the Hiroshima Memorial Peace Park: "The epitaph was written by Tadayoshi Saika, Professor of English Literature at Hiroshima University. He also provided the English translation, 'Let all the souls here rest in peace for we shall not repeat the evil.' On November 3, 1983, an explanation plaque in English was added in order to convey Professor Saika's intent that 'we' refers to 'all humanity,' not specifically the Japanese or Americans, and that the 'error' is the 'evil of war.'" Wikipedia, s.v. "Hiroshima Peace Memorial Park," last edited June 3, 2024, https://en.wikipedia.org/wiki/Hiroshima_Peace_Memorial_Park.
6 Ishiguro and Lee rarely appear together on the same literature syllabi. Ishiguro, who was born in Nagasaki, Japan, in 1954 and immigrated to Great Britain at age five, is seen as a key figure in global Anglophone literature, a field that has not only redrawn the postwar boundaries of British literary studies but also forced a reck-

oning with legacies of the British Empire insofar as many of its most prominent authors come from its former colonies and offer trenchant critiques of British colonialism in their writings. In contrast, Lee, who was born in Seoul, South Korea, in 1965 and immigrated to the United States at age three, is aligned with Asian American literature and US ethnic studies. These fields have introduced in parallel fashion histories of race, empire, immigration, and diaspora into a sanitized literary tradition of American exceptionalism, inclusion, and progress.

7 Freud, *Beyond the Pleasure Principle*, 9–10; emphasis in original.

8 Freud, *Beyond the Pleasure Principle*, 18; emphasis in original.

9 Freud, *Beyond the Pleasure Principle*, 22.

10 Freud, *Beyond the Pleasure Principle*, 22; emphasis in original.

11 Caruth, *Unclaimed Experience*, 2.

12 Caruth, *Unclaimed Experience*, 11; emphasis in original.

13 Caruth, *Unclaimed Experience*, 18.

14 Caruth, *Unclaimed Experience*, 2–3.

15 Caruth, *Unclaimed Experience*, 3.

16 Tasso, *La Gerusalemme liberata*, 162. I want to thank Matthew Aiello for drawing my attention to the emergence of the collective represented by Clorinda's voice in this passage. I want to thank David Young Kim for his assistance with the original Italian. The selection (from book 13, section 43) in Italian reads:

Clorinda fui, né sol qui spirto umano
albergo in questa pianta rozza e dura,
ma ciascun altro ancor, franco o pagano,
che lassi i membri a piè de l'alte mura,
astretto è qui da novo incanto e strano,
non so s'io dica in corpo o in sepoltura.
Son di sensi animati i rami e i tronchi,
e micidial sei tu, se legno tronchi.

The English translation from 1600 by Edward Fairfax reads:

I was Clorinda, now imprisoned here,
Yet not alone within this plant I dwell,
For every Pagan lord and Christian peer,
Before the city's walls last day that fell,
In bodies new or graves I wot not clear,
But here they are confined by magic's spell,
So that each tree hath life, and sense each bough,
A murderer if thou cut one twist art thou.

17 Caruth, *Unclaimed Experience*, 24.

18 Leys, *Trauma*, 268.

19 Leys, *Trauma*, 297.

20 See Craps, *Postcolonial Witnessing*, 15. Craps offers an overlapping observation about Caruth: "Given that this episode concerns the killing of an Ethiopian woman by a

European crusader, an orientalist dimension which Caruth does not acknowledge, her reading of this tale can be seen to illustrate the difficulty of trauma theory to recognize the experience of the non-Western other."

21 Rothberg, *Multidirectional Memory*, 90.

22 J. W. Scott, *On the Judgment of History*, xi.

23 Spivak, *Outside in the Teaching Machine*, 279.

24 Butler, *Giving an Account of Oneself*, 45.

25 Butler, *Frames of War*, 134–35.

26 In addition to Craps, *Postcolonial Witnessing*; and Rothberg, *Multidirectional Memory*, see Butler, *Parting Ways*.

27 J. W. Scott, *On the Judgment of History*, 1.

28 Conventional definitions of war crimes, the conduct of war under international law, held no provision for crimes committed by a sovereign power on its own citizens rather than on those of the enemy. As such, *crimes against humanity*, unlike the more narrow term *war crimes*, proliferated the subject and objects of violence. It encompassed not only acts of violence committed against civilian populations by other civilians during wartime or peacetime but also acts of violence that are either part of a systematic government policy or individual practice condoned by the government.

29 Craps, *Postcolonial Witnessing*, 3.

30 Writing about modern-day Israel, Joan Scott observes, "Jews were defined as potential victims in need of the protection of the Israeli security state. And, in an ironic twist, the achievement of their place in history came, for the Jewish victims of the Nazi genocide, in the form of an ethnically defined nation-state, which (as the quotation from Michael Mann I cited earlier maintains) rests on 'the notion that legitimate rule by *the people* over territory presupposed the absence (physically or culturally) of *other peoples* occupying that territory'" (J. W. Scott, *On the Judgment of History*, 20).

31 Derrida, "The Force of Law," 27.

32 See "Battleship Missouri Memorial," accessed September 7, 2020, https://ussmissouri.org/learn-the-history/world-war-ii-1#.

33 At the time of the bombing, Hiroshima's wartime population had been reduced to 250,000 from 380,000. Approximately 70,000–140,000 people were estimated to have perished from the atomic detonation in Hiroshima. A sprawling military-industrial complex, Hiroshima was a hub for Japan's enormous war efforts, the base of the notorious Fifth Infantry deployed to Southeast Asia during the war, and a center for the production of ships, munitions, and other matériel for the Imperial Army. The city also relied on conscripted labor from Japan's various colonial possessions in East and Southeast Asia. Had the Imperial Palace in Tokyo come under threat, Emperor Hirohito was to evacuate to Hiroshima.

34 Hersey, "Hiroshima," 15.

35 Hersey, "Hiroshima," 15.

36 Harold Ross, letter to E. B. White, quoted in Wikipedia, s.v. "*Hiroshima* (book)," last edited May 19, 2024, https://en.wikipedia.org/wiki/Hiroshima_(book)#cite_ref-New_7-0.

37 Steve Rothman, "The Publication of 'Hiroshima' in *The New Yorker*" (unpublished paper), January 8, 1997, accessed June 16, 2021, https://www.herseyhiroshima.com/hiro.php.

38 Wald, "Cells, Genes, and Stories," 4.

39 Hersey, "Hiroshima," 28.

40 Indeed, though he harbored little affection for the Japanese, Hersey himself remarked in the aftermath of the bombing that "if civilization was to mean anything, people had to acknowledge the humanity of their enemies." See Langewiesche, "The Reporter Who Told the World about the Bomb."

41 Klein, "Love, Guilt and Reparation," 66.

42 Dower, *War without Mercy*, 9.

43 Dower, *War without Mercy*, 11.

44 See "Fifth Air Force Weekly Intelligence Review," no. 86 (July 15–21, 1945), cited in Craven and Cate, *The Army Forces in World War II*, 696.

45 Hersey, "Hiroshima," 67–68.

46 See, for example, President Barack Obama's 2016 visit to the Hiroshima Peace Memorial Park, during which the president spoke at length about the costs of war and the need for peace and nuclear disarmament, but without apologizing for any US actions during World War II, including the use of atomic weapons.

47 Anghie, "Politic, Cautious, and Meticulous," 66. See also Dudden, *Troubled Apologies*. Dudden adds that the "chronic inability to confront how America's use of nuclear weapons against Japanese people in 1945 might constitute the kind of history for which survivors would seek an apology, let alone why the use of such weapons might represent a crime against humanity, is sustained by Washington's determination to maintain these weapons as the once and future legitimate tools of the national arsenal" (129). Were it otherwise, Dudden argues, "the likelihood that the use of atomic weapons on Japan would generate charges of attempted genocide and demands for apology and compensation against the United States or Harry Truman would increase exponentially" (129). Dudden continues that historians have "demonstrated through copious research on the bombs and the decisions to use them, Americans transferred what happened—the destruction of Hiroshima and Nagasaki—for an event that never took place—the proposed land invasion of Japan—to stand in for history. By the early 1950s, the imagined truth was American myth, and in 1959 President Truman wrote for the record that the bombs spared 'half a million' American lives, and that he 'never lost any sleep over the decision'" (118). Indeed, Hersey himself described the second bombing of Nagasaki as a criminal action.

48 See Schmitt, *The* Nomos *of the Earth*. It is ironic to note that, as Scott adds, "The London Charter listed aggressive warfare as one of the counts against the Nazis. It was signed by the Allies on August 8, 1945, the day [before] the United States bombed Nagasaki, two days after the bombing of Hiroshima; in February, the British and Americans had firebombed the city of Dresden—many thousands of civilians were killed in those raids. . . . Were these defensive operations or instances of aggressive warfare? And was it possible to insist that all war was a crime?" (J. W. Scott, *On the Judgment of History*, 12).

49 J. W. Scott, *On the Judgment of History*, xii.
50 Scheiber, "Taking Responsibility," 241.
51 Chen, *Asia as Method*, 7.
52 Yoneyama, "Traveling Memories, Contagious Justice," 66.
53 Dower, *Embracing Defeat*, 326.
54 Tanaka, McCormack, and Simpson, *Beyond Victor's Justice?*, xxix.
55 Dudden, *Troubled Apologies*, 37.
56 Scheiber, "Taking Responsibility," 245.
57 While the San Francisco Peace Treaty recognized that Japan should pay reparations, it also conceded that the Japanese government did not have the means to do so at the time, allowing the nation to prioritize its economic recovery over the needs of other nations. Japan eventually signed treaties with its East Asian neighbors to pay reparations. However, these payments largely took the form of loans to its former colonies such as South Korea, which in turn allowed Japanese corporations to profit.
58 Totani, *The Tokyo War Crimes Trial*, 13.
59 Totani, *The Tokyo War Crimes Trial*, 3.
60 A diplomatic missive from the Australian Department of External Affairs to the US government reveals the colonial priorities and racial stakes of the human at the IMTFE: "Many of the foulest atrocities in [Japanese] modern history [were] committed not only against the peoples of Eastern and Southeastern Asia but against nationals of Australia, the United States, and other Allied powers." Draft telegram to the Australian Embassy, Washington, DC, September 30, 1946, Department of External Affairs Records, A 1067/1, P 46/10/10/3, Australian National Archives, Canberra.
61 Totani adds, "The panel of judges was not so successful in nurturing the spirit of unity and mutual respect, either, as reflected in the split judgments between the majority opinion of eight judges, two separate concurrent opinions, and three separate dissenting opinions. This outcome stands in sharp contrast with the results of the Nuremberg trial, where the international judges overcame their differences and delivered a unanimous judgment" (*The Tokyo War Crimes Trial*, 12).
62 Gallicchio, *The Unpredictability of the Past*, 7. Gallicchio adds:

> A brief discussion of nomenclatures is in order. The varying perspectives that one can employ to view the subject is suggested by the terminology used to name it. The term "Asia-Pacific War" is a recent formulation. Japanese scholars seeking to emphasize Tokyo's aggression in Asia sometimes use the term "Fifteen-Year War" to describe Japan's wars in China, Southeast Asia, and the Pacific. Nationalists have countered with a revival of the term "Great East Asia War," a label that recalls Japanese wartime propaganda. On the other hand, official histories refer to the "Pacific War," a term first favored by American occupation authorities as an alternative to the politically charged "Greater East Asia War." But "Pacific War" also has its critics. They perceive in the label an attempt to justify Japanese actions by implying that Japan's real battle was against Western imperialism, in the form of the United States, and not China.

Official media in the People's Republic of China employ several labels, including "the Anti-Japanese war," for which the dates 1937 to 1945 are often used, and "the War of Resistance." In contrast, Americans tend to view the war against Japan as part of the larger struggle against fascism, most commonly known as World War II. When discussing the campaigns against Japan, American histories often focus on what the Americans call the Pacific Theatre, thereby appearing just like the war in China. As one can readily see, labels, even the most innocuous-sounding ones, reflect the user's assumptions and perspective. (7)

63 Gerry Simpson, "Writing the Tokyo Trial," in Tanaka, McCormack, and Simpson, *Beyond Victor's Justice?*, 27.
64 Takeshi Nakajima, "Justice Pal (India)," in Tanaka, McCormack, and Simpson, *Beyond Victor's Justice?*, 135.
65 Takeshi Nakajima, "Justice Pal (India)," in Tanaka, McCormack, and Simpson, *Beyond Victor's Justice?*, 134.
66 Yuki Tanaka, "The Atomic Bombing, the Tokyo Tribunal and the Shimoda Case: Lessons for Anti-nuclear Legal Movements," in Tanaka, McCormack, and Simpson, *Beyond Victor's Justice?*, 295. Pal writes, "It would be sufficient for my present purpose to say that if any indiscriminate destruction of civilian life and property is still illegitimate in warfare, then, in the Pacific war, this decision to use the atom bomb is the only near approach to the directives of the German Emperor during the first world war and of the Nazi leaders during the second world war. Nothing like this could be traced to the credit of the present accused."
67 Caruth, *Unclaimed Experience*, 3.
68 Butler, *Giving an Account of Oneself*, 37.
69 Butler, *Giving an Account of Oneself*, 45.
70 See Yoneyama, *Cold War Ruins*.
71 Dudden, *Troubled Apologies*, 37.
72 Ishiguro, *A Pale View of Hills*, 66. Yoneyama observes that pedagogical and political controversies

include the Ministry of Education's textbook approval system, which had until recently censored overt criticisms of Japan's military aggression and colonial policies; the repeated refusal of politicians to fully acknowledge the magnitude of the destruction caused by Japan; and the insufficiency of the government's undertakings in dealing with postwar reparations and colonial redress. In other words, there has been reflection among progressive citizens of Japan, who recognize that remembering a historical event only in terms of unprecedented self-victimization may serve to mystify other national conditions in the past and present. The Japanese citizen's counteramnesic practices have been underpinned by the conviction that to secure the memories of Japan's prewar and wartime imperialism and military aggression in school textbooks and other public apparatuses that produce national history is inseparably linked to pursuit of peace, human rights, and democratic ideals of a civil society. Moreover, many construe the act of remembering Japan's military and colonial pasts as leading also to critically reflecting on

Japan's postwar neocolonial economic dominance in the region. ("For Transformative Knowledge," 334)

73 See Tansman, *The Cultures of Japanese Fascism*. Tansman writes, "After 1937, producers of newspapers, radio broadcasts, books, songs, comic books, films, and photographs were all subjected to strict codes, or subjected themselves, out of pragmatic necessity, to self-censorship. Left-wing political organizations and journals were squelched, and language thought to deter the war effort and the unity of a citizenry behind it came under harsh review" (13).
74 Ishiguro, *An Artist of the Floating World*, 123, 124, 134.
75 Ishiguro, *An Artist of the Floating World*, 123.
76 Klein, "Love, Guilt and Reparation," 104–5.
77 Ishiguro, *An Artist of the Floating World*, 191.
78 Ishiguro, *An Artist of the Floating World*, 192–93.
79 "I tend to be attracted to pre-war and postwar settings," Ishiguro notes in an interview by Graham Swift in BOMB *Magazine*, "because I'm interested in this business of values and ideals being tested, and people having to face up to the notion that their ideals weren't quite what they thought they were before the test came." See Swift, "Kazuo Ishiguro."
80 Dudden, *Troubled Apologies*, 36.
81 Yoneyama, "Traveling Memories, Contagious Justice," 65.
82 Lie, *Zainichi (Koreans in Japan)*, x. Even after the American occupation of Japan ended, internal discrimination against Zainichi persisted. According to Lie, the 1952 San Francisco Peace Treaty, which officially ended the American occupation of Japan, continued to function on the exclusion of Koreans from Japan, "restor[ing] Japanese sovereignty but rescind[ing] Japanese citizenship for ethnic Koreans remaining in Japan."
83 C. Lee, *A Gesture Life*, 73. See also Cheng, "Passing, Natural Selection, and Love's Failure," 559. In her reading of Hata's transformation, Cheng notes that what is astonishing in his renunciation of his ethnic identification for this national identification "is not the insight that assimilation for a disenfranchised person can provide material advantages but rather its revelation into the unseen, ontological dividends afforded by such assimilation." For Hata, in other words, to become Japanese is to become human, as to become American is to become a citizen. See also Jerng, "Recognizing the Transracial Adoptee." Relatedly, Jerng foregrounds the ambivalent identification Hata sustains with Koreanness—with K in Burma as well as with Sunny in the United States—through the trope of adoption and passing.
84 Cheng, "Passing, Natural Selection, and Love's Failure," 561.
85 Nguyen, *The Gift of Freedom*.
86 Yoneyama, "Traveling Memories, Contagious Justice," 82.
87 Malcolm X, *The Autobiography of Malcolm X*, 179.
88 See chapter 3 of Moyn, *The Last Utopia*.
89 See Borstelmann, *The Cold War and the Color Line*; Dudziak, *Cold War Civil Rights*; and Westad, *The Global Cold War*.

90 See A. Y. Lee, "Repairing Police Action after the Korean War in Toni Morrison's *Home*"; and Kim, *The Interrogation Rooms of the Korean War*.
91 See Ngai, *Impossible Subjects*, 175–201.
92 House Resolution 422, August 10, 1988.
93 Cho, *Haunting the Korean Diaspora*, 8.
94 Dunant, *A Memory of Solferino*.
95 C. Lee, *The Surrendered*, 416.
96 A. Y. Lee, "Someone Else's War," 173.
97 Gourevitch, "Alms Dealers," 106.
98 Hopgood, *The Endtimes of Human Rights*, x.
99 Pietz, "The 'Post-colonialism' of Cold War Discourse," 57–58.
100 See Cheah, "Crises of Money."

3. BEYOND SOVEREIGNTY

1 See Wikipedia, s.v. "Uranium mining," last edited July 19, 2024, https://en.wikipedia.org/wiki/Uranium_mining. During World War II, uranium supplies were mined mainly in the Belgian Congo, Canada, and the United States.
2 See Bothwell, *Eldorado*. Bothwell recounts that Gilbert and Charlie LaBine, brothers from Ottawa and founders of the Eldorado Gold Mines Company in Manitoba in 1926 (later the Eldorado Mining and Refining Limited Company), began mining at Port Radium in Echo Bay, Great Bear Lake, in 1934. At the time, radium was considered a miracle cure for cancer, selling at $70,000 per gram. It was also used for medical X-rays as well as for glow-in-the-dark watch dials that allowed military personnel to read time in the dark. Many radium painters developed disfiguring cancers such as radium jaw.

Radium's discovery broke the world monopoly on radium held by a Belgian mining company in the Congo. In response, the Belgian company lowered the price of radium to $25,000 per gram. The LaBine enterprise almost went bankrupt. When scientists in Europe and North American discovered the immense energy potential released by splitting the atom in the late 1930s, the wartime demand for uranium surged. In 1942, the United States contracted to buy all the uranium produced by the Eldorado Mine. When the Canadian government secretly nationalized the enterprise, the LaBine brothers became instant millionaires.

The operating dates of the mine vary slightly according to different sources. Cindy Kenny-Gilday recounts that the site operated as a radium mine from 1934 to 1939, a uranium mine from 1942 to 1962, and a silver mine from 1962 to 1982. See "Impacts of Uranium Mining at Port Radium, NWT, Canada," last updated September 7, 2005, accessed February 4, 2024, https://www.wise-uranium.org/uippra.html.

Reporter Andrew Nikiforuk writes that the site operated as a radium mine from 1932 to 1940 and a uranium mine from 1941 to 1960. See Nikiforuk, "Echoes of the Atomic Age: Cancer Kills Fourteen Aboriginal Uranium Workers."
3 See Bothwell, *Eldorado*; and Nikiforuk, "Echoes of the Atomic Age: Cancer Kills Fourteen Aboriginal Uranium Workers."

4 Clements, *Burning Vision*, 90.
5 Horkheimer and Adorno, *Dialectic of Enlightenment*, 3; my emphasis. The linking of radiation with atomic disaster is not coincidental. The English translation reads, "In the most general sense of progressive thought, the Enlightenment has always aimed at liberating men from fear and establishing their sovereignty. Yet the fully enlightened earth *radiates* disaster triumphant."
6 Masco, *The Nuclear Borderlands*, 34.
7 See Nixon, *Slow Violence and the Environmentalism of the Poor*.
8 Derrida, *On Cosmopolitanism and Forgiveness*, 59.
9 For great acceleration, see Nixon, "The Great Acceleration and the Great Divergence." See also Chakrabarty, "The Climate of History," 206–7. Chakrabarty writes:

> To call human beings geological agents is to scale up our imagination of the human. Humans are biological agents, both collectively and as individuals. They have always been so. There was no point in human history when humans were not biological agents. But we can become geological agents only historically and collectively, that is, when we have reached numbers and invented technologies that are on a scale large enough to have an impact on the planet itself. To call ourselves geological agents is to attribute to us a force on the same scale as that released at other times when there has been a mass extinction of species. We seem to be currently going through that kind of a period. The current "rate in the loss of species diversity," specialists argue, "is similar in intensity to the event around 65 million years ago which wiped out the dinosaurs." Our footprint was not always that large. Humans began to acquire this agency only since the Industrial Revolution, but the process really picked up in the second half of the twentieth century.

10 Masco, "The Age of Fallout," 137.
11 Masco, "The Age of Fallout," 137.
12 See Kuletz, *The Tainted Desert*.
13 The December 1999 arrest and (largely) failed prosecution of Wen Ho Lee, a Taiwan-born scientist working at LANL accused by the US government of espionage of behalf of the People's Republic of China (PRC), is an enduring symptom of the competing anxieties of national security and international dependency as well as anti-Asian sentiment in the West. Lee was charged in a fifty-nine-count indictment that eventually fell apart. After federal investigators were unable to prove their initial accusations, the government conducted a separate investigation and was ultimately able to charge Lee only with improper handling of restricted data, one of the original fifty-nine indictment counts, to which he pleaded guilty as part of a plea settlement. In June 2006, Lee received $1.6 million from the federal government and five media organizations as part of a settlement of a civil suit he had filed against them for leaking his name to the press before any formal charges had been filed against him. Federal judge James A. Parker eventually apologized to Lee for denying him bail and putting him in solitary confinement, and excoriated the US government for misconduct and misrepresentations to the court. See Wikipedia, s.v. "Wen Ho Lee," last edited June 9, 2024, http://en.wikipedia.org/wiki

/Wen_Ho_Lee. See also Masco, *The Nuclear Borderlands*, 273. Masco points out that during the Manhattan Project it was European physicists—Enrico Fermi, Leo Szilard, Hans Bethe, Edward Teller, and others—who made the project a success. In the 1980s, furthermore, both Los Alamos and Lawrence Livermore Laboratories had directors who were naturalized US citizens.

14 See Gimbel, *Science, Technology, and Reparations*. German nuclear facilities were also bombarded and destroyed at the end of World War II so that nuclear technologies could not be seized by the Soviet Union.

15 Laduke, *All Our Relations*, 98.

16 See Schwartz, *Atomic Audit*.

17 Israel seems to be the lone exception here, although the country will not admit to either having or not having nuclear arms. Pre-apartheid South Africa also had nuclear arms. They were dismantled after the fall of apartheid. See Wikipedia, s.v. "Nuclear Proliferation," last edited July 20, 2024, http://en.wikipedia.org/wiki/Nuclear_proliferation.

18 Anghie, "Politic, Cautious, and Meticulous," 66.

19 Voyles, *Wastelanding*, 3. Henningson, *Somba Ke*, is a follow-up documentary to *A Village of Widows* that chronicles the reopening of the Eldorado Mine at Port Radium, Echo Bay.

20 Anghie, "Politic, Cautious, and Meticulous," 66; emphasis in original.

21 Masco, *The Nuclear Borderlands*, 299.

22 See Bryan-Wilson, "Building a Marker of Nuclear Warning," 184.

23 Masco, "The Age of Fallout," 153, 138.

24 Hersey, "Hiroshima," 49–50.

25 Masco, *The Nuclear Borderlands*, 33.

26 Chakrabarty, "Postcolonial Studies and the Challenge of Climate Change," 2. See also Chakrabarty, "The Climate of History." Invoking Walter Benjamin's melancholic Angel of History, Chakrabarty emphasizes that "species may indeed be the name of a placeholder for an emergent, new universal history of humans that flashes up in the moment of the danger that is climate change" (221). Chakrabarty's slide into anthropomorphism underscores the need for scholars to engage more systematically with the burgeoning field of environmental racism and to analyze the human and nonhuman dimensions of climate change together with the insights of postcolonial and critical race theory.

27 There is a growing body of cultural production on the toxic colonialisms of these interconnected spaces: from documentaries such as Blow's *A Village of Widows*, Henningson's *Somba Ke: The Money Place*, and Spitz's *The Return of Navajo Boy* to novels such as Leslie Marmon Silko's *Ceremony* and Gerald Vizenor's *Hiroshima Bugi: Atomu 57*. According to *Navajo Boy*, more than one thousand uranium mines have been built since the start of the Cold War on Navajo lands, resulting in similar effects as the Sahtu Dene. Mining ceased on Navajo lands in 1923 and was transferred to Belgian Congo. These activities resumed after World War II. Miners who worked in uranium-filled mines have a very high incidence of cancer relative to the rest of the United States population. Though the Navajo workers and families noticed this

in the 1950s, bureaucrats dragged their feet, and companies disregarded warnings. The miners, especially the Navajo miners, were kept from receiving compensation for the suffering they went through. Only in 1990 was a law passed known as the Radiation Exposure Compensation Act (RECA).

28 See Nixon, "The Great Acceleration and the Great Divergence." Nixon emphasizes that "the most influential Anthropocene intellectuals have sidestepped the question of unequal human agency, unequal human impacts, and unequal human vulnerabilities."

29 Masco, *The Nuclear Borderlands*, 118, 108, 36.

30 Masco, *The Nuclear Borderlands*, 140.

31 Voyles, *Wastelanding*, 4.

32 See Hämäläinen, *The Comanche Empire*. Hämäläinen writes about this phenomenon of what I call "reverse visual hallucination"—of refusing to see what appears to be seen in the barren Western landscape of Manifest Destiny. His analysis echoes, I suggest, the Doctrine of Discovery and legal principles of *terra nullius* ("nobody's land") in international law justifying the colonial settlement and acquisition of Australia, which remained the law of the land until the landmark 1992 ruling by the High Court of Australia in *Mabo v. Queensland*.

33 See Wikipedia, s.v. "British Nuclear Tests at Maralinga," last edited July 16, 2024, http://en.wikipedia.org/wiki/British_nuclear_tests_at_Maralinga. Between 1955 and 1963, the British government conducted several major nuclear detonations as well as hundreds of minor nuclear tests on the lands of the Maralinga Tjarutja, a southern Pitjantjatjara aborginal people. The poisoning of the region has led to two massive cleanup efforts in 1967 and 2000 and reparations of $13.5 million from the British to the Australian government in 1994. Not surprisingly, these funds have not been provided directly to the Tjarutja people to compensate for the debilitating environmental and health effects of the nuclear testing on their lands; rather, they have been applied to the cleanup efforts in the region. Furthermore, in yet another manifestation of sovereign authority, the Australian government instead manages these funds on behalf of those aboriginal groups suffering from radiation poisoning and illness, while indemnifying Britain from further legal liability.

34 Anghie, "Politic, Cautious, and Meticulous," 62; brackets in original.

35 *Cherokee Nation v. State of Georgia*, 12. In the US legal academy, almost no Native American cases are taught as part of the standard law curriculum. If *Cherokee* is presented, it is not in Constitutional Law but rather in Civil Procedure, for as Marshall's opinion so clearly states, the question is not whether the Cherokee have rights but whether they have standing to sue the state of Georgia in US federal court. They of course do not. On the "trinity" of Marshall's Native American cases in nineteenth-century law—*Johnson v. McIntosh* (1823), *Cherokee Nation v. State of Georgia* (1931), and *Worcester v. Georgia* (1932)—see Barker, "For Whom Sovereignty Matters."

36 Cunsolo and Ellis, "Ecological Grief," 275.

37 See Cunsolo and Ellis, "Ecological Grief," 278. Writing about Australia and Canada, Cunsolo and Ellis observe, "Our research indicates that many individuals identi-

fied feeling anticipatory grief for ecological changes that had not yet happened. In these cases, grief for anticipated future ecological losses is also tied to grief over future losses to culture, livelihoods and ways of life (see also the film *Attutauniujut Nunami/Lament for the Land*). A similar form of anticipatory grieving has also been documented amongst Sami reindeer herders in Northern Sweden fearful of the disappearance of their valued way of life" (278). They continue:

> First, ecological grief is likely to be more common amongst peoples that retain close living and working relationships to natural environments than those who do not. Second, different types of climate hazards may elicit different experiences of ecological grief, with the effects of gradual and cumulative climate change less discernible than those associated with acute weather-related disasters. Third, people living in areas where high climate risk maps onto existing and entrenched vulnerabilities are more likely to experience ecological grief than people living in areas of low climate risk and low vulnerability. Finally, personal and cultural notions of value are likely to underpin grief responses, so that the intensity of ecological grief experienced is proportional to the value attributed to the ecological loss. (279)

38 Blow, *A Village of Widows*, at 1:09.
39 Blow, *A Village of Widows*, at 17:47.
40 I would like to thank Mariam Banahi, who corresponded with me about my reading of the film at Johns Hopkins in October 2018, for this critical insight. Email from October 11, 2018.
41 See Nikiforuk, "Echoes of the Atomic Age: Cancer Kills Fourteen Aboriginal Uranium Workers." See also Pasternak, *Yellow Dirt*.
42 Blow, *A Village of Widows*, at 25:00.
43 Blow, *A Village of Widows*, at 28:40.
44 See Nikiforuk, "Echoes of the Atomic Age: Uranium Haunts a Northern Aboriginal Village." According to Nikiforuk, Canada's Department of Mines knew of the health dangers of radon gas and radioactive dust in Port Radium as early as 1932. In 1949, US officials raised health concerns about Port Radium miners, but this report was deemed confidential and "not to be quoted in any published report."
45 See Nikiforuk, "Echoes of the Atomic Age: Uranium Haunts a Northern Aboriginal Village."
46 See Henningson, *Somba Ke*.
47 Picq, "Self-Determination as Anti-extractivism," 4.
48 Blow, *A Village of Widows*, at 5:01.
49 Derrida, *On Cosmopolitanism and Forgiveness*, 32.
50 Derrida, *On Cosmopolitanism and Forgiveness*, 31.
51 On apology as apologia, see Barkan and Karn, "Group Apology as an Ethical Imperative," 17.
52 *Cherokee Nation v. State of Georgia*, 10.
53 See D'Cruz, *Identity Politics in Deconstruction*, 89. D'Cruz has an excellent take on Derrida's "secret" of forgiveness, writing, "There is a distinction here between *judgment* in the juridical process that is concerned with the calculability of the law and

the *decision* to forgive, which is inaccessible to the law. . . . The secret of the dividing line between the decision to forgive or not to forgive remains" (91). Indeed, forgiveness, if unconditional, neither demands nor solicits apology and repair. It is proffered without obligation, expectation, or exchange. One can forgive without apology, and one can apologize without forgiveness.

54 See Saldaña-Portillo, *Indian Given*. I am referring more to Anglo settler colonies and strategies of dispossession in Canada and the United States. As Saldaña-Portillo observes, Spanish settlement in Mexico functioned on the division of the civilized and savage Indian, who could be either included or excluded from the settler state through conversion and mixing (*mestizaje*).

55 Derrida, *On Cosmopolitanism and Forgiveness*, 32, 39.

56 Caruth, *Unclaimed Experience*, 24.

57 In the age of the Anthropocene, we need to move away from notions of trauma as individual, temporal, and linguistic to notions of trauma as collective, spatial, and material. It thus also demands a reconsideration of conventional approaches to trauma as a singular, shattering event, while evoking "an understanding of humanist history and subjectivity that displaces (without entirely eliminating) the positions of victim and perpetrator." See Rothberg, "Preface: Beyond Tancred and Clorinda," xvi. See also Gone, "Colonial Genocide and History Trauma in Native North America." Gone also argues against notions of victim and perpetrator in regard to dominant historical narratives of Indigenous genocide that reduces heterogenous histories to one narrative of elimination. He points out that an intransigent narrative of colonial perpetrator and Indigenous victim works to limit strategies of reparation and reconciliation across these fixed divisions.

58 See Butler, *Parting Ways*, 43. I derive this reading from Butler. Commenting on Emmanuel Levinas, she writes:

> We do not take responsibility for the Other's suffering only when it is clear that we have caused that suffering. In other words, we do not take responsibility only for the clear choices we have made and the effects they have had. Although, of course, such acts are important components of any account of responsibility, they do not indicate its most fundamental structure. According to Levinas, we affirm the unfreedom at the heart of our relations with others, and only by ceding in this way do we come to understand responsibility. In other words, I cannot disavow my relation to the Other, regardless of what the other does, regardless of what I might will. Indeed, responsibility is not a matter of cultivating a will (as it is for Kantians), but of recognizing an unwilled susceptibility as a resource for being responsive to the Other. (43)

59 See Levinas, *Otherwise Than Being*.

60 Van Wyck, "An Emphatic Geography," 172.

61 Butler, *Giving an Account of Oneself*, 136.

62 See Henningson, *Somba Ke*.

63 Blow, *A Village of Widows*, at 41:36.

64 See work by Barker, Simpson, Kevin Bruyneel, Jodi Byrd, Glen Sean Coulthard, and J. Kēhaulani Kauanui.

65 See Byrd, "Indigenous Futures beyond the Sovereignty Debate."

66 See Alfred, *Peace, Power, Righteousness*; Deloria, "Intellectual Self-Determination and Sovereignty"; Morris, "International Law and Politics"; and Warrior, *Tribal Secrets*. See also Ford, "Locating Indigenous Self-Determination," 2–3. Ford writes:

> The exercise of settler jurisdiction over indigenous people remains patchy, and evolving definitions of indigenous governance and indigenous land rights by settler courts constantly redefine the relationship among sovereignty, territory and jurisdiction. The contemporary relationship of indigenous rights to land (dominium) and to autonomy or sovereignty (imperium) has yet to be resolved by philosophers and lawyers. Their uncertain relationship is evident in shifting Supreme Court definitions of the province of federal, state and Indian jurisdiction in the United States that have increasingly attenuated the capacity of long-established indigenous governments to govern Indian reservations. Meanwhile, the growing value of indigenous land claims in remote parts of Australia and Canada has proved even more challenging. On the one hand, indigenous land claims before settler courts have been predicated on ancient association, sacralized possession and corporate identity; even at their weakest, they subtly affirm indigenous corporate autonomy. On the other hand, mining booms and the ecological turn have transformed economic "wastelands" into important sources of revenue, material bases that could be used to support much stronger institutions of governance among indigenous communities. Canadian and Australian courts in particular have yet to reconcile common law notions of "property" with claims by Aborigines that they should be able to "speak for country," regulate visitors, or negotiate with mining companies about access to mineral rights which have mostly been reserved to the Crown. (2–3)

67 Coulthard, *Red Skin, White Masks*, 12.

68 Coulthard, *Red Skin, White Masks*, 12–13.

69 Blow, *A Village of Widows*, at 35:49.

70 See J. C. Scott, *The Art of Not Being Governed*. Like Zomia's geographies, the Dene's legal status as Canadian subjects overlaps with other political orders and priorities.

71 See Yoneyama, *Hiroshima Traces*, esp. chap. 5, "Ethnic and Colonial Memories: The Korean Atom Bomb Memorial." Yoneyama writes in *Hiroshima Traces* that

> under the National Manpower Mobilization Act (as extended to Koreans between 1939 and 1945) the Japanese government brought 700,000 Koreans to Japan for forced labor in coal mines, in munitions factories, and at various other dangerous construction sites. Because many Koreans worked in the factories located near Hiroshima, the city's Korean population also increased toward the end of the war. Yet these facts about the relationship between Korea and Hiroshima have not been widely known. The Korean victims and their specific sufferings have been virtually absent from past official representations of Hiroshima's atomic atrocity, which portray the devastation in ways that encourage memories of a universal human existence. . . . Until 1990, the speeches of political elites at the annual municipal Peace Memorial Ceremony on 6 August never referred to the 20,000 to 30,000 (and per-

haps even more) Korean atom bomb dead, who comprised between 10 and 20 percent of those killed immediately in the Hiroshima bombing. (152)

In a personal correspondence, Yoneyama tells me,

> It seems highly likely that the hospital... is the Kawamura Clinic. It is a hospital not necessarily devoted to Korean hibakusha as mentioned in [the documentary], but accommodates mostly Korean but all overseas survivors who hold the official *hibakusha* certificate but whose medical expenses are not subsidized by the Japanese government because of their residence abroad. Over decades Kawamura clinic has supported the citizens' activism to bring overseas *hibakusha* (most of whom are from Korea and Brazil) to be examined and treated there so that their medical expenses can be covered under the *Hibakusha* medical law which used to apply only to those hibakusha residing in Japan. In other words, before the law changed in 2001 only those with the official *hibakusha* certificate residing in Japan (regardless of nationality, thanks to the earlier activism), could qualify for such subsidy.... The key activist behind this volunteer work was school teacher Toyonaga Keisaburo, whose essay we included in *Perilous Memories*.

72 Blow, *A Village of Widows*, at 43:15.
73 Ivy, "Trauma's Two Times," 172.
74 As I discussed in chapter 2 of this book, Hiroshima was a major military industrial complex. It produced both arms and ships for the war effort, and it served as the base for the Fifth Infantry Army deployed to Nanking and Burma during World War II.
75 Byrd, *The Transit of Empire*, 200.
76 Byrd, *The Transit of Empire*, xxiii–xxiv.
77 Derrida, "Nietzsche and the Machine," 48.
78 Yoneyama, *Hiroshima Traces*, 3.
79 Byrd, "What's Normative Got to Do with It?"
80 See Serres, *The Natural Contract*. They insist on an account of what philosopher Serres describes as a "natural" rather than "social" contract with the planet, one that brings reciprocity and balance in our relations to others as well as to the lands that sustain us and exceed our comprehension and existence.
81 D'Cruz, *Identity Politics in Deconstruction*, 84.
82 See Spivak, *Death of a Discipline*. I borrow this concept of a "non-coercive rearrangement of desire" from Spivak.

Bibliography

Alfred, Taiaiake. *Peace, Power, Righteousness: An Indigenous Manifesto*. Toronto: Oxford University Press, 1999.

Ames, James Barr. *Lectures on Legal History and Miscellaneous Legal Essays*. Cambridge, MA: Harvard University Press, 1913.

Anghie, Antony. "Francisco de Vitoria and the Colonial Origins of International Law." *Social and Legal Studies* 5, no. 3 (1996): 321–36.

Anghie, Antony T. "Politic, Cautious, and Meticulous: An Introduction to the Symposium on the Marshall Islands Case." *American Society of International Law* 111 (2017): 62–67. https://doi.org/10.1017/aju.2017.27.

Arendt, Hannah. *The Origins of Totalitarianism*. 1951. Reprint, New York: Harcourt Brace Jovanovich, 1968.

Armitage, David. "John Locke, Carolina, and the 'Two Treatises of Government.'" *Political Theory* 32, no. 5 (October 2004): 602–27.

Barkan, Elazar, and Alexander Karn. "Group Apology as an Ethical Imperative." In *Taking Wrongs Seriously: Apologies and Reconciliation*, edited by Barkan and Karn, 3–30. Stanford, CA: Stanford University Press, 2006.

Barker, Joanne. "For Whom Sovereignty Matters." In *Sovereignty Matters: Locations of Contestation and Possibility in Indigenous Struggles for Self-Determination*, edited by Joanne Barker, 1–31. Lincoln: University of Nebraska Press, 2005.

Berlant, Lauren. *Cruel Optimism*. Durham, NC: Duke University Press, 2011.

Bickel, Alexander M. *The Morality of Consent*. New Haven, CT: Yale University Press, 1975.

Blow, Peter, dir. *A Village of Widows*. Toronto: Kinetic Video, 1999.

Borstelmann, Thomas. *The Cold War and the Color Line: American Race Relations in the Global Arena*. Cambridge, MA: Harvard University Press, 2003.

Bothwell, Robert. *Eldorado: Canada's National Uranium Company*. Toronto: University of Toronto Press, 1984.

Brown, Wendy. *States of Injury: Power and Freedom in Late Modernity*. Princeton, NJ: Princeton University Press, 1995.

Bruyneel, Kevin. *The Third Space of Sovereignty: The Postcolonial Politics of U.S.-Indigenous Relations*. Minneapolis: University of Minnesota Press, 2007.

Bryan-Wilson, Julia. "Building a Marker of Nuclear Warning." In *Monument and Memory, Made and Unmade*, edited by Robert S. Nelson and Margaret Olin, 183–204. Chicago: University of Chicago Press, 2004.

Butler, Judith. *Frames of War: When Is Life Grievable?* New York: Verso, 2009.

Butler, Judith. *Giving an Account of Oneself*. New York: Fordham University Press, 2005.

Butler, Judith. "Moral Sadism and Doubting One's Own Love." In *Reading Melanie Klein*, edited by Lyndsey Stonebridge and John Phillips, 179–89. London: Routledge, 1998.

Butler, Judith. *Parting Ways: Jewishness and the Critique of Zionism*. New York: Columbia University Press, 2012.

Byrd, Jodi A. "Indigenous Futures beyond the Sovereignty Debate." In *The Cambridge History of Native American Literature, Part IV—Visions and Revisions: 21st-Century Prospects*, edited by Melanie Benson Taylor, 501–18. Cambridge: Cambridge University Press, 2020.

Byrd, Jodi A. *The Transit of Empire: Indigenous Critiques of Colonialism*. Minneapolis: University of Minnesota Press, 2011.

Byrd, Jodi A. "What's Normative Got to Do with It? Toward Indigenous Queer Relationality." *Social Text* 38, no. 4 (2020): 105–23.

Caruth, Cathy. *Empirical Truths and Critical Fictions: Locke, Wordsworth, Kant, Freud*. Baltimore, MD: Johns Hopkins University Press, 1991.

Caruth, Cathy. *Unclaimed Experience: Trauma, Narrative, and History*. Baltimore, MD: Johns Hopkins University Press, 1996.

Césaire, Aimé. *Discourse on Colonialism*. Translated by Joan Pinkham. New York: Monthly Review Press, 2001.

Chakrabarty, Dipesh. "The Climate of History: Four Theses." *Critical Inquiry* 35, no. 2 (2009): 197–222.

Chakrabarty, Dipesh. "Postcolonial Studies and the Challenge of Climate Change." *New Literary History* 43, no. 1 (Winter 2012): 1–18.

Chakrabarty, Dipesh. *Provincializing Europe: Postcolonial Thought and Historical Difference*. Princeton, NJ: Princeton University Press, 2000.

Chambers-Letson, Joshua. "Reparative Feminisms, Repairing Feminism—Reparation, Postcolonial Violence, and Feminism." *Women and Performance: A Journal of Feminist Theory* 16, no. 2 (July 2006): 169–89.

Chang, Iris. *The Rape of Nanking: The Forgotten Holocaust of World War II*. New York: Penguin, 1997.

Cheah, Pheng. "Crises of Money." *positions* 16, no. 1 (2008): 189–219.

Chen, Kuan-hsing. *Asia as Method: Toward Deimperialization*. Durham, NC: Duke University Press, 2010.

Cheng, Anne. "Passing, Natural Selection, and Love's Failure: Ethics of Survival from Chang-rae Lee to Jacques Lacan." *American Literary History* 17, no. 3 (2005): 553–74.

Cho, Grace. *Haunting the Korean Diaspora: Shame, Secrecy, and the Forgotten War*. Minneapolis: University of Minnesota Press, 2008.

Clements, Marie. *Burning Vision*. Vancouver: Talon, 2003.

Cohen, Ed. *A Body Worth Defending: Immunity, Biopolitics, and the Apotheosis of the Modern Body*. Durham, NC: Duke University Press, 2009.
Coulthard, Glen Sean. *Red Skin, White Masks: Rejecting the Colonial Politics of Recognition*. Minneapolis: University of Minnesota Press, 2014.
Craps, Stef. *Postcolonial Witnessing: Trauma Out of Bounds*. New York: Palgrave Macmillan, 2013.
Craven, W. F., and J. L. Cate, eds. *The Army Forces in World War II*. Vol. 5. Chicago: University of Chicago Press, 1948.
Cunsolo, Ashlee, and Neville R. Ellis. "Ecological Grief as a Mental Health Response to Climate Change–Related Loss." *Nature Climate Change* 8 (April 2018): 275–81.
D'Cruz, Carolyn. *Identity Politics in Deconstruction: Calculating the Incalculable*. London: Ashgate, 2008.
Deloria, Vine, Jr. "Intellectual Self-Determination and Sovereignty: Looking at the Windmills in Our Minds." *Wicazo Sa Review* 13, no. 1 (1998): 25–31.
Derrida, Jacques. "The Force of Law." In *Deconstruction and the Possibility of Justice*, edited by Drucilla Cornell, Michel Rosenfeld, and David Gray Carlson, 3–67. New York: Routledge, 1992.
Derrida, Jacques. "Nietzsche and the Machine—Interview with Jacques Derrida by Richard Beardsmith." *Journal of Nietzsche Studies* 7 (1994): 7–66.
Derrida, Jacques. *On Cosmopolitanism and Forgiveness*. Translated by Mark Dooley and Michael Hughes. New York: Routledge, 1991.
Douzinas, Costas. *Human Rights and Empire: The Political Philosophy of Cosmopolitanism*. London: Routledge, 2007.
Dower, John W. *Embracing Defeat: Japan in the Wake of World War II*. New York: New Press, 1999.
Dower, John W. *War without Mercy: Race and Power in the Transpacific*. New York: Pantheon, 1986.
Dubois, Laurent. *Haiti: The Aftershocks of History*. New York: Picador, 2013.
Du Bois, W. E. B. *The World and Africa*. New York: International Publishers, 1979.
Dudden, Alexis. *Troubled Apologies among Japan, Korea, and the United States*. New York: Columbia University Press, 2008.
Dudziak, Mary. *Cold War Civil Rights: Race and the Image of American Democracy*. Princeton, NJ: Princeton University Press, 2011.
Dunant, Jean-Henri. *A Memory of Solferino*. English ed. Geneva: International Committee of the Red Cross, 1939.
Fanon, Frantz. *Black Skin, White Masks*. 1952. Reprint, New York: Grove, 2008.
Fanon, Frantz. *The Wretched of the Earth*. 1962. Reprint, New York: Grove, 2021.
Ford, Lisa. "Locating Indigenous Self-Determination in the Margins of Settler Sovereignty: An Introduction." In *Between Indigenous and Settler Governance*, edited by Ford and Tim Rowse, 1–11. London: Routledge, 2013.
Foucault, Michel. *The Care of the Self*. Translated by Graham Burchell. New York: Pantheon, 1986.
Foucault, Michel. *Hermeneutics of the Subject*. Translated by Graham Burchell. New York: Palgrave Macmillan, 2005.

Freud, Sigmund. *Beyond the Pleasure Principle* (1920). In *The Standard Edition of the Complete Psychological Works of Sigmund Freud*, vol. 18: *1920–1922*, translated and edited by James Strachey et al., 1–64. London: Hogarth, 1959.

Freud, Sigmund. *Mourning and Melancholia* (1917). In *The Standard Edition of the Complete Psychological Works of Sigmund Freud*, vol. 14: *1914–1916*, translated and edited by James Strachey et al., 243–58. London: Hogarth, 1959.

Freud, Sigmund. *Thoughts for the Times on War and Death* (1915). In *The Standard Edition of the Complete Psychological Works of Sigmund Freud*, vol. 14: *1914–1916*, translated and edited by James Strachey et al., 273–300. London: Hogarth, 1959.

Gallicchio, Marc, ed. *The Unpredictability of the Past: Memories of the Asia-Pacific War in U.S.-East Asian Relations*. Durham, NC: Duke University Press, 2007.

Gay, Peter. *Freud: A Life for Our Time*. New York: W. W. Norton, 1988.

Gerwarth, Robert, and Stephan Malinowski. "Hannah Arendt's Ghosts: Reflections on the Disputable Path from Windhoek to Auschwitz." *Central European History* 42 (2009): 279–300.

Gimbel, John. *Science, Technology, and Reparations: Exploitation and Plunder in Postwar Germany*. Stanford, CA: Stanford University Press, 1990.

Gone, Joseph P. "Colonial Genocide and History Trauma in Native North America: Complicating Contemporary Attributions." In *Colonial Genocide in Indigenous North America*, edited by Jeff Benvenuto, Andrew Woolford, and Alexander Laban Hinton, 273–91. Durham, NC: Duke University Press, 2014.

Gourevitch, Philip. "Alms Dealers: Can You Provide Humanitarian Aid without Facilitating Conflicts?" *New Yorker*, October 11, 2010, 102–9.

Hämäläinen, Pekka. *The Comanche Empire*. New Haven, CT: Yale University Press, 2008.

Henningson, David, dir. *Somba Ke: The Money Place*. New York: Filmakers Library, 2007.

Hersey, John. "Hiroshima." *New Yorker*, August 31, 1946, 15–68.

Hopgood, Stephen. *The Endtimes of Human Rights*. Ithaca, NY: Cornell University Press, 2013.

Horkheimer, Max, and Theodor W. Adorno. *Dialectic of Enlightenment*. Translated by John Cumming. New York: Continuum, 1999.

Hull, Isabel V. *Absolute Destruction: Military Culture and the Practices of War in Imperial Germany*. Ithaca, NY: Cornell University Press, 2006.

Hunt, Lynn. *Inventing Human Rights: A History*. New York: W. W. Norton, 2008.

Ishiguro, Kazuo. *An Artist of the Floating World*. New York: Vintage, 1989.

Ishiguro, Kazuo. *A Pale View of Hills*. New York: Vintage, 1990.

Ishiguro, Kazuo. *The Remains of the Day*. New York: Vintage, 1990.

Ivy, Marilyn. "Trauma's Two Times: Japanese Wars and Postwars." *positions* 16, no. 1 (2008): 165–88.

Jackson, Zakiyyah Iman. *Becoming Human: Matter and Meaning in an Antiblack World*. New York: New York University Press, 2020.

Jefferies, Matthew. *Contesting the German Empire, 1871–1918*. Malden, MA: Wiley-Blackwell, 2008.

Jerng, Mark C. "Recognizing the Transracial Adoptee: Adoption Life Stories and Chang-rae Lee's *A Gesture Life*." MELUS 31, no. 2 (2006): 41–67.

Kang, Laura. *The Traffic in Asian Women*. Durham, NC: Duke University Press, 2020.
Kantorowicz, Ernest H. *The King's Two Bodies*. Princeton, NJ: Princeton University Press, 1997.
Kauanui, J. Kēhaulani. *Paradoxes of Hawaiian Sovereignty: Land, Sex, and the Colonial Politics of State Nationalism*. Durham, NC: Duke University Press, 2018.
Kim, Monica. *The Interrogation Rooms of the Korean War: The Untold History*. Princeton, NJ: Princeton University Press, 2019.
Klein, Melanie. "A Contribution to the Psychogenesis of Manic-Depressive States" (1935). In *The Selected Melanie Klein*, edited by Juliet Mitchell, 115–45. New York: Free Press, 1986.
Klein, Melanie. "Love, Guilt and Reparation" (1937). In *Love, Hate and Reparation*, 57–119. New York: W. W. Norton, 1964.
Klein, Melanie. "The Psycho-analytic Play Technique: Its History and Its Significance" (1955). In *The Selected Melanie Klein*, edited by Juliet Mitchell, 35–54. New York: Free Press, 1986.
Kuletz, Valerie. *The Tainted Desert: Environmental and Social Ruin in the American West*. New York: Routledge, 1998.
Laduke, Winona. *All Our Relations: Native Struggles for Land and Life*. Boston: South End, 1999.
Langewiesche, William. "The Reporter Who Told the World about the Bomb." *New York Times*, August 4, 2020. https://www.nytimes.com/2020/08/04/books/review/fallout-hiroshima-hersey-lesley-m-m-blume.html.
Laslett, Peter. "Introduction." In *Two Treatises of Government*, edited by Peter Laslett, 3–16. Cambridge: Cambridge University Press, 1988.
Latour, Bruno. *We Have Never Been Modern*. Translated by Catherine Porter. Cambridge, MA: Harvard University Press, 1993.
Laubender, Carolyn. "Beyond Repair: Interpretation, Reparation, and Melanie Klein's Clinical Play-Technique." *Studies in Gender and Sexuality* 20, no. 1 (2019): 51–67.
Lazenby, J. F. *The First Punic War*. New York: Routledge, 1996.
Lee, A. Yumi. "Repairing Police Action after the Korean War in Toni Morrison's *Home*." *Radical History Review* 137 (May 2020): 119–39.
Lee, A. Yumi. "Someone Else's War: Korea and the Post-1945 Racial Order." PhD diss., University of Pennsylvania, 2015.
Lee, Chang-rae. *A Gesture Life*. New York: Riverhead, 1999.
Lee, Chang-rae. *The Surrendered*. New York: Riverhead, 2010.
Levinas, Emmanuel. *Otherwise Than Being, or Beyond Essence*. Translated by Alphonso Lingis. The Hague, Netherlands: Martinus Nijhoff, 1981.
Leys, Ruth. *Trauma: A Genealogy*. Chicago: University of Chicago Press, 2000.
Lie, John. *Zainichi (Koreans in Japan): Diasporic Nationalism and Postcolonial Identity*. Berkeley: University of California Press, 2008.
Lindqvist, Sven. *"Exterminate All the Brutes": One Man's Odyssey into the Heart of Darkness and the Origins of European Genocide*. Translated by Joan Tate. New York: New Press, 2007.
Locke, John. *An Essay concerning Human Understanding*. Edited by Peter H. Nidditch. New York: Oxford University Press, 1979.

Locke, John. *Two Treatises of Government*. Edited by Peter Laslett. Cambridge: Cambridge University Press, 1988.

Lowe, Lisa. "The Intimacies of Four Continents." In *Haunted by Empire: Geographies of Intimacy in North American History*, edited by Ann Laura Stoler, 191–212. Durham. NC: Duke University Press, 2006.

Madley, Benjamin. "From Africa to Auschwitz: How German Southwest Africa Incubated Ideas and Methods Adopted and Developed by the Nazis in Eastern Europe." *European History Quarterly* 3, no. 5 (2005): 429–64.

Malcolm X. *The Autobiography of Malcolm X: As Told to Alex Haley*. New York: Ballantine, 1992.

Mannoni, Octave. *Psychology of Colonization*. Ann Arbor: University of Michigan Press, 1991.

Masco, Joseph. "The Age of Fallout." *History of the Present* 5, no. 2 (2015): 137–68.

Masco, Joseph. *The Nuclear Borderlands: The Manhattan Project in Post–Cold War New Mexico*. Princeton, NJ: Princeton University Press, 2006.

McCarthy, Mary. "The Hiroshima *New Yorker*." *Politics*, November 1946, 367.

Mehta, Uday S. "Liberal Strategies of Exclusion." *Politics and Society* 18, no. 4 (1990): 427–54.

Mills, Charles W. *The Racial Contract*. Ithaca, NY: Cornell University Press, 1999.

Minow, Martha. "Breaking the Cycles of Hatred." In *Breaking the Cycles of Hatred: Memory, Law, and Repair*, edited by Martha Minow and Nancy L. Rosenbaum, 14–76. Princeton, NJ: Princeton University Press, 2002.

Moon, Katharine H. S. "Military Prostitution and the U.S. Military in Asia." *Asia-Pacific Journal* 7, no. 3 (2009): 1–10.

Morris, Glenn T. "International Law and Politics: Toward a Right for Self-Determination for Indigenous Peoples." In *The State of Native America: Genocide, Colonization, and Resistance*, edited by M. Annette Jaimes, 55–86. Boston: South End, 1992.

Moses, A. Dirk. *Colonialism and Genocide*. New York: Routledge, 1996.

Moyn, Samuel. *The Last Utopia: Human Rights in History*. Cambridge, MA: Harvard University Press, 2010.

Musser, Amber. *Sensational Flesh: Race, Power, and Masochism*. New York: New York University Press, 2014.

Ngai, Mae M. *Impossible Subjects: Illegal Aliens and the Making of Modern America*. Princeton, NJ: Princeton University Press, 2004.

Nguyen, Mimi Thi. *The Gift of Freedom: War, Debt, and Other Refugee Passages*. Durham, NC: Duke University Press, 2012.

Nichols, Robert. *Theft Is Property! Dispossession and Critical Theory*. Durham, NC: Duke University Press, 2020.

Nikiforuk, Andrew. "Echoes of the Atomic Age: Cancer Kills Fourteen Aboriginal Uranium Workers." *Calgary Herald*, March 14, 1998, 1. https://www.ccnr.org/deline_deaths.html.

Nikiforuk, Andrew. "Echoes of the Atomic Age: Uranium Haunts a Northern Aboriginal Village." *Calgary Herald*, March 14, 1998, A4. https://www.ccnr.org/deline_deaths.html.

Nixon, Rob. "The Great Acceleration and the Great Divergence: Vulnerability in the Anthropocene." *PMLA Profession*, March 19, 2014. http://profession.commons.mla.org/2014/03/19/the-great-acceleration-and-the-great-divergence-vulnerability-in-the-anthropocene/.

Nixon, Rob. *Slow Violence and the Environmentalism of the Poor*. Cambridge, MA: Harvard University Press, 2013.

O'Brien, Jean. *Firsting and Lasting: Writing Indians Out of Existence in New England*. Minneapolis: University of Minnesota Press, 2010.

Olusoga, David, and Caspar W. Erichsen. *The Kaiser's Holocaust: Germany's Forgotten Genocide and the Colonial Roots of Nazism*. London: Faber & Faber, 2010.

Pasternak, Judy. *Yellow Dirt: A Poisoned Land and the Betrayal of the Navajos*. New York: Free Press, 2010.

Perugini, Nicola, and Neve Gordon. *The Human Right to Dominate*. New York: Oxford University Press, 2015.

Phillips, John. "Editor's Note." In *Reading Melanie Klein*, edited by Lyndsey Stonebridge and John Phillips, 126–27. London: Routledge, 1998.

Picq, Manuela Lavinas. "Self-Determination as Anti-extractivism: How Indigenous Resistance Challenges IR." *E-International Relations*, May 21, 2014, 1–6. http://www.e-ir.info/2014/05/21/self-determination-as-anti-extractivism-how-indigenous-resistance-challenges-ir/.

Pietz, William. "The 'Post-colonialism' of Cold War Discourse." *Social Text* 19/20 (1988): 55–75.

Pitts, Jennifer. *A Turn to Empire: The Rise of Imperial Liberalism in Britain and France*. Princeton, NJ: Princeton University Press, 2005.

Poiger, Uta G. *Jazz, Rock, and Rebels: Cold War Politics and American Culture in a Divided Germany*. Berkeley: University of California Press, 2000.

Riviere, Joan. "On the Genesis of Psychical Conflict in Earliest Infancy." *International Journal of Psychoanalysis* 17 (1936): 395–422.

Rose, Jacqueline. *Why War? Psychoanalysis, Politics, and the Return to Klein*. Oxford: Blackwell, 1993.

Rothberg, Michael. *The Implicated Subject*. Stanford, CA: Stanford University Press, 2019.

Rothberg, Michael. *Multidirectional Memory: Remembering the Holocaust in the Age of Decolonization*. Stanford, CA: Stanford University Press, 2009.

Rothberg, Michael. "Preface: Beyond Tancred and Clorinda: Trauma Studies for Implicated Subjects." In *The Future of Trauma Theory: Contemporary Literary and Cultural Criticism*, edited by Gert Beulens, Sam Durrant, and Robert Eaglestone, xi–xviii. London: Routledge, 2014.

Ruskola, Teemu. "Canton Is Not Boston: The Invention of American Imperial Sovereignty." *American Quarterly* 57, no. 3 (2005): 859–84.

Saldaña-Portillo, María-Josefina. *Indian Given: Racial Geographies across Mexico and the United States*. Durham, NC: Duke University Press, 2016.

Sarkin, Jeremy. *Germany's Genocide of the Herero: Kaiser Wilhelm II, His General, His Settlers, His Soldiers*. Martlesham, UK: James Curry, 2011.

Scheiber, Harry N. "Taking Responsibility: Moral and Historical Perspectives on the

Japanese War-Reparations Issues." *Berkeley Journal of International Law* 20, no. 1 (2002): 233–49.
Schmitt, Carl. *The* Nomos *of the Earth in the International Law of the* Jus Publicum Europaeum. Translated by G. L. Ulmen. New York: Telos, 2003.
Schumann, Dirk. *Political Violence in the Weimar Republic, 1918–1933: Battle for the Streets and Fears of Civil War*. Oxford: Berghahn, 2009.
Schwartz, Stephen, ed. *Atomic Audit: The Costs and Consequences of U.S. Nuclear Weapons since 1940*. Washington, DC: Brookings Institution Press, 1998.
Scott, James C. *The Art of Not Being Governed: An Anarchist History of Upland Southeast Asia*. New Haven, CT: Yale University Press, 2010.
Scott, Joan Wallach. *On the Judgment of History*. New York: Columbia University Press, 2021.
Sedgwick, Eve Kosofsky. "Melanie Klein and the Difference Affect Makes." *South Atlantic Quarterly* 106, no. 3 (2007): 625–42.
Sedgwick, Eve Kosofsky. *Touching Feeling: Affect, Pedagogy, Performativity*. Durham, NC: Duke University Press, 2003.
Segel, Hanna. *Introduction to the Work of Melanie Klein*. New York: Karnac, 1988.
Seitz, David K. "A Wizard of Disquietude in Our Midst: Melanie Klein and the Critical Geographies of Manic Reparation." *Society and Space* 41, no. 2 (2023): 1–19.
Serres, Michel. *The Natural Contract: Meditations on Environmental Change and the Necessity of a Pact between Earth and Its Inhabitants*. Translated by Elizabeth MacArthur and William Paulson. Ann Arbor: University of Michigan Press, 1995.
Shapira, Michal. *The War Inside: Psychoanalysis, Total War, and the Making of the Democratic Self in Postwar Britain*. Cambridge: Cambridge University Press, 2013.
Silko, Leslie Marmon. *Ceremony*. New York: Penguin, 2006.
Spelman, Elizabeth V. *Repair: The Impulse to Restore in a Fragile World*. Boston: Beacon, 2003.
Spillers, Hortense. "Mama's Baby, Papa's Maybe: An American Grammar Book." *Diacritics* 17, no. 2 (1987): 64–81.
Spitz, Jeff, dir. *The Return of Navajo Boy*. 2000. DVD, 15th anniv. ed. Chicago: Groundswell Educational Films, 2011.
Spivak, Gayatri Chakravorty. *Death of a Discipline*. New York: Columbia University Press, 2003.
Spivak, Gayatri Chakravorty. *Outside in the Teaching Machine*. New York: Routledge, 1993.
Stacey, Jackie. "Wishing Away Ambivalence." *Feminist Theory* 15, no. 1 (2014): 39–49.
Swift, Graham. "Kazuo Ishiguro." *BOMB Magazine*, Fall 1989. http://bombsite.com/issues/29/articles/1269.
Tanaka, Yuki, Tim McCormack, and Gerry Simpson, eds. *Beyond Victor's Justice? The Tokyo War Crimes Trial Revisited*. Leiden, Netherlands: Brill, 2011.
Tansman, Alan, ed. *The Cultures of Japanese Fascism*. Durham, NC: Duke University Press, 2009.
Tasso, Torquato. *Jerusalem Delivered.* Translated by Edward Fairfax. London, 1600. Project Gutenberg. https://www.gutenberg.org/ebooks/392.
Tasso, Torquato. *La Gerusalemme liberate*. Edited by Pietro Papini. Florence, Italy: Sansoni, 1917.

Thucydides. *The History of the Peloponnesian War*. Translated by Rex Warner. New York: Penguin Classics, 1972.
Totani, Yuma. *The Tokyo War Crimes Trial: The Pursuit of Justice in the Wake of World War II*. Cambridge, MA: Harvard University Asia Center, 2008.
Traverso, Enzo. *The Origins of Nazi Violence*. Translated by Janet Lloyd. New York: New Press, 2003.
Trouillet, Michel-Rolph. *Silencing the Past: Power and the Production of History*. Boston: Beacon, 2015.
Tully, James. *An Approach to Political Philosophy: Locke in Contexts*. Cambridge: Cambridge University Press, 1993.
van Wyck, Peter C. "An Emphatic Geography: Notes on the Ethical Itinerary of Landscape." *Canadian Journal of Communication* 33 (2008): 171–91.
Vizenor, Gerald. *Hiroshima Bugi: Atomu 57*. Lincoln: University of Nebraska Press, 2010.
Voyles, Traci Brynne. *Wastelanding: Legacies of Uranium Mining in Navajo Country*. Minneapolis: University of Minnesota Press, 2015.
Wald, Priscilla. "Cells, Genes, and Stories: Human Being under the Microscope in the Aftermath of War." Paper presented at the Ward-Phillips Lecture Series "Human Being after Genocide," University of Notre Dame, Notre Dame, IN, April 1, 2009.
Wald, Priscilla. "Exquisite Fragility: Human Being in the Aftermath of War." In *A Companion to American Literary Studies*, edited by Caroline F. Levander and Robert S. Levine, 437–53. Malden, MA: Blackwell, 2011.
Warrior, Robert Allen. *Tribal Secrets: Recovering American Indian Intellectual Traditions*. Minneapolis: University of Minnesota Press, 1995.
Weheliye, Alexander G. *Habeas Viscus: Racializing Assemblages, Biopolitics, and Black Feminist Theories of the Human*. Durham, NC: Duke University Press, 2014.
Weitz, Eric D. *A Century of Genocide: Utopias of Race and Nation*. Princeton, NJ: Princeton University Press, 2005.
Wellerstein, Alex. "Counting the Dead at Hiroshima and Nagasaki." *Bulletin of the Atomic Scientists*, August 4, 2020. https://thebulletin.org/2020/08/counting-the-dead-at-hiroshima-and-nagasaki/.
Westad, Odd Arne. *The Global Cold War*. Cambridge: Cambridge University Press, 2007.
Williams, Randall. *The Divided World: Human Rights and Its Violence*. Minneapolis: University of Minnesota Press, 2010.
Wynter, Sylvia. "Unsettling the Coloniality of Being/Power/Truth/Freedom: Towards the Human, After Man, Its Overrepresentation—an Argument." *New Centennial Review* 3, no. 3 (2003): 257–337.
Yoneyama, Lisa. *Cold War Ruins: Transpacific Critique of American Justice and Japanese War Crimes*. Durham, NC: Duke University Press, 2016.
Yoneyama, Lisa. "For Transformative Knowledge and Postnationalist Public Spheres: The Smithsonian *Enola Gay* Controversy." In *Perilous Memories: The Asia-Pacific War(s)*, edited by T. Fujitani, Geoffrey M. White, and Lisa Yoneyama, 323–46. Durham, NC: Duke University Press, 2001.
Yoneyama, Lisa. *Hiroshima Traces: Time, Space, and the Dialectics of Memory*. Berkeley: University of California Press, 1999.

Yoneyama, Lisa. "Traveling Memories, Contagious Justice: Americanization of Japanese War Crimes at the End of the Post–Cold War." *Journal of Asian American Studies* 6, no. 1 (2003): 57–93.

Zaretsky, Eli. "Melanie Klein and the Emergence of Modern Personal Life." In *Reading Melanie Klein*, edited by Lyndsey Stonebridge and John Phillips, 32–50. London: Routledge, 1998.

Index

Note: Page references in italics indicate figures.